Math Mammoth
Grade 7-A Worktext

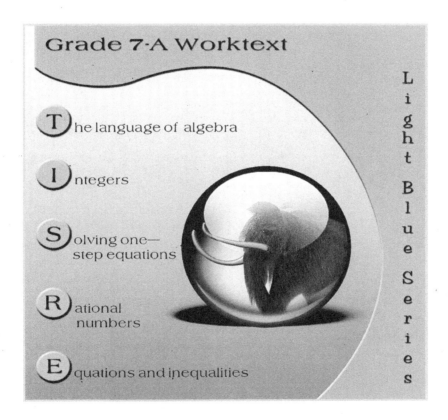

By Maria Miller

Copyright 2014 - 2023 Taina Maria Miller
ISBN 978-1511766241

2014 Edition

All rights reserved. No part of this book may be reproduced or transmitted in any form or by any means, electronic or mechanical, or by any information storage and retrieval system, without permission in writing from the author.

Copying permission: For having purchased this book, the copyright owner grants to the teacher-purchaser a limited permission to reproduce this material for use with his or her students. In other words, the teacher-purchaser MAY make copies of the pages, or an electronic copy of the PDF file, and provide them at no cost to the students he or she is actually teaching, but not to students of other teachers. This permission also extends to the spouse of the purchaser, for the purpose of providing copies for the children in the same family. Sharing the file with anyone else, whether via the Internet or other media, is strictly prohibited.

No permission is granted for resale of the material.

The copyright holder also grants permission to the purchaser to make electronic copies of the material for back-up purposes.

If you have other needs, such as licensing for a school or tutoring center, please contact the author at
https://www.MathMammoth.com/contact

Contents

Foreword 5

Chapter 1: The Language of Algebra
Introduction 6
The Order of Operations 11
Expressions and Equations 15
Properties of the Four Operations 18
Simplifying Expressions 22
Growing Patterns 1 26
The Distributive Property 29
Chapter 1 Review 34

Chapter 2: Integers
Introduction 36
Integers 41
Addition and Subtraction on the Number Line 45
Addition of Integers 49
Subtraction of Integers 52
Adding or Subtracting Several Integers 56
Distance and More Practice 58
Multiplying Integers 62
Dividing Integers 67
Negative Fractions 69
The Order of Operations 73
Chapter 2 Mixed Review 75
Chapter 2 Review 77

Chapter 3: Solving One–Step Equations
Introduction 80
Solving Equations 82
Addition and Subtraction Equations 89
Multiplication and Division Equations 93

Word Problems 97
Constant Speed 100
Chapter 3 Mixed Review 107
Chapter 3 Review 110

Chapter 4: Rational Numbers

Introduction 112
Rational Numbers 116
Adding and Subtracting Rational Numbers 124
Multiply and Divide Rational Numbers 1 130
Multiply and Divide Rational Numbers 2 134
Many Operations with Rational Numbers 141
Scientific Notation 145
Equations with Fractions 148
Equations with Decimals 153
Chapter 4 Mixed Review 156
Chapter 4 Review 158

Chapter 5: Equations and Inequalities

Introduction 163
Two-Step Equations 168
Two-Step Equations: Practice 173
Growing Patterns 2 177
A Variable on Both Sides 181
Some Problem Solving 187
Using the Distributive Property 190
Word Problems 196
Inequalities 199
Word Problems and Inequalities 204
Graphing 206
An Introduction to Slope 210
Speed, Time and Distance 215
Chapter 5 Mixed Review 220
Chapter 5 Review 223

Foreword

Math Mammoth Grade 7 comprises a complete math curriculum for the seventh grade mathematics studies. This is a pre-algebra course, so students can continue to an algebra 1 curriculum after studying this.

The curriculum meets and actually exceeds the Common Core Standards (CCS) for grade 7. The two major areas where it exceeds those standards are linear equations (chapter 5) and the Pythagorean Theorem (chapter 9). Linear equations are covered in more depth than the CCS requires, and the Pythagorean Theorem belongs to grade 8 in the CCS. You can access a document detailing the alignment information either on the Math Mammoth website or in the download version of this curriculum.

The main areas of study in Math Mammoth Grade 7 are:

- The basics of algebra (expressions, equations, inequalities, graphing);
- Integers;
- Ratios, proportions, and percent;
- Geometry;
- Probability and statistics.

This book, 7-A, covers the language of algebra (chapter 1), integers (chapter 2), one-step equations (chapter 3), rational numbers (chapter 4), and equations and inequalities (chapter 5). The rest of the topics are covered in the 7-B worktext.

Some important points to keep in mind when using the curriculum:

- The two books (parts A and B) are like a "framework", but you still have a lot of liberty in planning your student's studies. The five chapters in part 7-A are best studied in the order presented. However, you can study the chapters on geometry, probability, and statistics at most any point during the year. The chapters on ratios & proportions and percent (in part 7-B) are best left until the student has learned to solve one-step equations (in chapter 3).

- Math Mammoth is mastery-based, which means it concentrates on a few major topics at a time, in order to study them in depth. However, you can still use it in a *spiral* manner, if you prefer. Simply have your student study in 2-3 chapters simultaneously. This type of flexible use of the curriculum enables you to truly individualize the instruction for the student.

- Don't automatically assign all the exercises. Use your judgment, trying to assign just enough for your student's needs. You can use the skipped exercises later for review. For most students, I recommend to start out by assigning about half of the available exercises. Adjust as necessary.

- For review, the curriculum includes a worksheet maker (Internet access required), mixed review lessons, additional cumulative review lessons, and the word problems continually require usage of past concepts. Please see more information about review (and other topics) in the FAQ at
  https://www.mathmammoth.com/faq-lightblue.php

I heartily recommend that you view the full user guide for your grade level, available at
https://www.mathmammoth.com/userguides/

Lastly, you can find free videos matched to the curriculum at https://www.mathmammoth.com/videos/

   I wish you success in teaching math!

   Maria Miller, the author

Chapter 1: The Language of Algebra
Introduction

In the first chapter of *Math Mammoth Grade 7* we review all of the sixth grade algebra topics and also study some basic properties of the operations.

The main topics are the order of operations, expressions, and simplifying expressions in several different ways. The main principles are explained and practiced both with visual models and in abstract form, and the lessons contain varying practice problems that approach the concepts from various angles.

This chapter is like an introduction that lays a foundation for the rest of the year. For example, when we study integers in the next chapter, students will once again simplify expressions, just with negative numbers. Then when we study equations in chapters 3 and 5, students will again simplify expressions, use the distributive property, and solve equations.

Please note that it is not recommended to assign all the exercises by default. Use your judgment, and strive to vary the number of assigned exercises according to the student's needs. See the user guide at https://www.mathmammoth.com/userguides/ for some further thoughts on using and pacing the curriculum.

You can find matching videos for topics in this chapter at https://www.mathmammoth.com/videos/ (choose grade 7).

The Lessons in Chapter 1

	page	span
The Order of Operations	11	*4 pages*
Expressions and Equations	15	*3 pages*
Properties of the Four Operations	18	*4 pages*
Simplifying Expressions	22	*4 pages*
Growing Patterns 1	26	*3 pages*
The Distributive Property	29	*5 pages*
Chapter 1 Review	34	*2 pages*

Helpful Resources on the Internet

You can also access this list of links at https://links.mathmammoth.com/gr7ch1

ORDER OF OPERATIONS

Otter Rush
Practice exponents in this otter-themed math game.
https://www.mathplayground.com/ASB_Otter_Rush.html

Exponents Jeopardy Game
Practice evaluating exponents, equations with exponents, and exponents with fractional bases in this interactive Jeopardy-style game.
https://www.math-play.com/Exponents-Jeopardy/exponents-jeopardy-math-game_html5.html

Choose A Math Operation
Choose the mathematical operation(s) so that the number sentence is true. Practice the role of zero and one in basic operations or operations with negative numbers. Helps develop number sense and logical thinking.
https://www.homeschoolmath.net/operation-game.php

Order of Operations Quiz
A 10-question online quiz that includes two different operations and possibly parentheses in each question. You can also modify the quiz parameters yourself.
https://www.thatquiz.org/tq-1/?-j8f-lk-p0

The Order of Operations Millionaire
Answer multiple-choice questions that have to do with the order of operations, and win a million. Can be played alone or in two teams.
https://cutt.ly/Order-of-Operations-Millionaire

Exploring Order of Operations (Object Interactive)
The program shows an expression, and you click on the correct operation (either +, −, ×, ÷ or exponent) to be done first. The program then solves that operation, and you click on the *next* operation to be performed, *etc.*, until it is solved. Lastly, the resource includes a game where you click on the falling blocks in the sequence that the order of operations would dictate. Note: May load slowly.
https://www.learnalberta.ca/content/mejhm/html/object_interactives/order_of_operations/use_it.html

Make a Number Game
Arrange the number cards and the operation symbols, so that the expression will make the target number.
https://www.mathplayground.com/make_a_number.html

Order of Operations Practice
A simple online quiz of 10 questions. Uses parentheses and the four operations.
https://www.onlinemathlearning.com/order-of-operations-practice.html

Make and Take
Practice the order of operations with this simple card game.
https://mathhombre.blogspot.com/2011/11/make-and-take.html

WRITING EXPRESSIONS

Writing Basic Expressions with Variables
Practice writing expressions with this interactive online quiz.
https://www.khanacademy.org/math/algebra-basics/alg-basics-algebraic-expressions/alg-basics-writing-expressions/e/writing-expressions-with-variables-1

Who Wants to Be a Hundredaire?
Try to work your way up to $100 by responding correctly to multiple-choice questions about algebraic expressions.
https://bit.ly/Who-Wants-to-Be-a-Hundredaire

Equivalent Algebraic Expressions
Practice determining whether or not two algebraic expressions are equivalent by manipulating the expressions. These problems require you to combine like terms and apply the distributive property.
https://cutt.ly/Equivalent-Algebraic-Fractions

Visual Patterns
Click on the pattern to see a larger image and the answer to step 43. Can you solve the equation?
https://www.visualpatterns.org

Equations of Sequence Patterns
An instructional video from Khan Academy.
https://www.youtube.com/watch?v=_3BnyEr5fG4

Write Equations for Non-Linear Patterns
Learn how to break down complex shapes that grow in 2 dimensions into smaller terms, making it easy to write an equation and find values for graphing.
https://www.youtube.com/watch?v=ecARQzwvN9w

Expressions and Variables Quiz
Choose an equation to match the word problem or situation.
https://www.softschools.com/quizzes/math/expressions_and_variables/quiz815.html

Translating Words to Algebraic Expressions
Match the correct math expression with the corresponding English phrase, such as "7 less than a number". You can do this activity either as a matching game or as a concentration game.
https://www.quia.com/jg/1452190.html

Rags to Riches - Verbal and Algebraic Expressions
Translate between verbal and algebraic expressions in this quest for fame and fortune.
https://www.quia.com/rr/520475.html

Algebra Noodle
Play a board game against the computer while modeling and solving simple equations and evaluating simple expressions. Choose level 2 (level 1 is too easy for 7th grade). Note: May load slowly.
https://www.free-training-tutorial.com/math-games/algebra-noodle.html

Matching Algebraic Expressions with Word Phrases
Five sets of word phrases to match with expressions.
https://bit.ly/Matching-Algebraic-Expressions-with-Word-Phrases

PROPERTIES OF THE OPERATIONS

Properties of Operations at Quizlet
Includes explanations, online flashcards, and a test for the properties of operations (commutative, associate, distributive, inverse, and identity properties). The inverse and identity properties are not covered in this chapter of Math Mammoth but can be learned at the website. The identity property refers to the special numbers that do not change addition or multiplication results (0 and 1).
https://quizlet.com/2799611/properties-of-operation-flash-cards/

Commutative/Associative/Distributive Properties Matching Game
Match the terms and expressions in the two columns.
https://www.quia.com/cm/61114.html?AP_rand=1554068841

Properties of Multiplication
Simple online practice about the commutative, associative, distributive, and identity properties of multiplication.
https://www.aaamath.com/pro74b-propertiesmult.html

Properties of Multiplication
Simple online practice about the commutative, associative, distributive, and identity properties of multiplication.
https://www.aaamath.com/pro74ax2.htm

Properties of the Operations Scatter Game
Drag the corresponding items to each other to make them disappear.
https://quizlet.com/763838/scatter

Associative, Distributive and Commutative Properties
Examples of the various properties followed by a simple self-test.
https://bit.ly/Associative-Distributive-and-Commutative-Properties

SIMPLIFYING EXPRESSIONS

Simplifying Algebraic Expressions Quiz
An online quiz of 15 questions.
https://www.quia.com/quiz/1200540.html

Distributive Property Quiz
Reinforce your skills with this interactive online quiz.
https://www.thatquiz.org/tq/practicetest?13y7ctojwrlbs

Algebra Pairs
Practice simplifying expressions with these interactive online activities.
https://cutt.ly/Algebra-Pairs

Combining Like Terms
Combine the like terms to create an equivalent expression.
https://bit.ly/combine-like-terms

THE DISTRIBUTIVE PROPERTY

Factor the Expressions Quiz
Factor expressions such as $3x + 15$ into $3(x + 5)$.
https://www.thatquiz.org/tq-0/?-jh00-l3-p0

Brackets
Expand algebraic expressions containing brackets and simplify the resulting expression.
https://www.transum.org/software/SW/Starter_of_the_day/Students/Brackets.asp?Level=4

Distributive Property Game
Solve questions related to the usage of the distributive property amidst playing a game. Play either a bouncing balls game or free-kick soccer game, with the same questions.
https://reviewgamezone.com/games3/bounce.php?test_id=22828&title=DISTRIBUTIVE%20PROPERTY

https://bit.ly/Distributive-Property-Game

EVALUATE EXPRESSIONS

Evaluating Expressions Quiz
Includes ten multiple-choice questions.
https://maisonetmath.com/algebra/algebra-quizzes/290-evaluating-expressions

Evaluating Expressions with Multiple Variables
Use this interactive online exercise for additional practice.
https://cutt.ly/evaluate-expressions-multiple-variables

TERMINOLOGY

How to Identify Terms, Like Terms, Coefficients, and Constants
A detailed explanation about how to identify the parts of an expression.
https://socratic.org/questions/how-do-you-identify-the-terms-like-terms-coefficients-and-constants-in-each-expr-1

Identifying Variable Parts and Coefficients of Terms
After the explanations, you can generate exercises by pushing the button that says "new problem." The script shows you a multiplication expression, such as $-(3e)(3z)m$, and you need to identify its coefficient and variable part, effectively by first simplifying it.
https://www.onemathematicalcat.org/algebra_book/online_problems/id_var_part_coeff.htm#exercises

Identifying Parts of Algebraic Expressions
Answer questions about variables, expressions, equations, and inequalities in this multiple-choice test.
https://cutt.ly/Identifying-Parts-of-Algebraic-Expressions

Algebra - Basic Definitions
Clear definitions with illustrations of basic algebra terminology, including term, coefficient, constant, and expression.
https://www.mathsisfun.com/algebra/definitions.html

GENERAL

Balance Beam Activity
A virtual balance that poses puzzles where the student must think algebraically to find the weights of various figures. Includes three levels.
https://mste.illinois.edu/users/pavel/java/balance/index.html

Nine Digits Puzzles
Practice your reasoning skills with these interactive online puzzles.
https://www.transum.org/software/SW/Starter_of_the_day/students/hot/NineDigits.asp?Level=2

Algebraic Expressions - Online Assessment
During this online quiz you must simplify expressions, combine like terms, use the distributive property, express word problems as algebraic expressions and recognize when expressions are equivalent. Each incorrect response will allow you to view a video explanation for that problem.
https://maisonetmath.com/algebra/algebra-quizzes/287-algebraic-expressions-assessment

The Order of Operations

Let's review! Exponents are a shorthand for writing repeated multiplications by the same number.

For example, $0.9 \cdot 0.9 \cdot 0.9 \cdot 0.9 \cdot 0.9$ is written 0.9^5.

The tiny raised number is called the **exponent**.
It tells us how many times the **base** number is multiplied by itself.

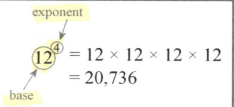

The expression 2^5 is read as "two to the fifth power," "two to the fifth," or "two raised to the fifth power."

Similarly, 0.7^8 is read as "seven tenths to the eighth power" or "zero point seven to the eighth."

The "powers of 6" are simply expressions where 6 is raised to some power: for example, 6^3, 6^4, 6^{45}, and 6^{99} are powers of 6.

Expressions with the exponent 2 are usually read as something "squared." For example, 11^2 is read as "eleven squared." That is because it gives us the area of a square with sides 11 units long.

Similarly, if the exponent is 3, the expression is usually read using the word "cubed." For example, 1.5^3 is read as "one point five cubed" because it is the volume of a cube with an edge 1.5 units long.

1. Evaluate.

   a. 4^3 **64**
   b. 10^5 **100,000**
   c. 0.1^2 **0.01**
   d. 0.2^3 **0.008**
   e. 1^{100} **1**
   f. 100 cubed **1,000,000**

2. Write these expressions using exponents. Find their values.

   a. $0 \cdot 0 \cdot 0 \cdot 0 \cdot 0$ $0^5 = 0$
   b. $0.9 \cdot 0.9$ $0.9^2 = 0.81$
   c. $5 \cdot 5 \cdot 5 + 2 \cdot 2 \cdot 2 \cdot 2 \cdot 2$ $5^3 + 2^5 = 157$; $125 + 32$
   d. $6 \cdot 10 \cdot 10 \cdot 10 \cdot 10 \cdot 10 \cdot 10 - 9 \cdot 10 \cdot 10 \cdot 10 \cdot 10 = 5,910,000$

The expression $(5\text{ m})^3$ means that we multiply 5 meters by itself three times:

$5\text{ m} \cdot 5\text{ m} \cdot 5\text{ m} = 125\text{ m}^3$ 5 meters (unit of measure)

Notice that $(5\text{ m})^3$ is different from 5 m^3. The latter has no parentheses, so the exponent (the little 3) applies only to the unit "m" and not to the whole quantity 5 m.

(varible)

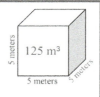

3. Find the value of the expressions.

   a. $(2\text{ cm})^3$ **8 cm³**
   b. $(11\text{ ft})^2$ **121 ft²**
   c. $(1.2\text{ km})^2$ **1.44 km²**

4. Which expression from the right matches with (a) and (b) below?

   a. The volume of a cube with edges 2 cm long.
   b. The volume of a cube with edges 8 cm long.

   (i) 8 cm^3 (ii) $(8\text{ cm})^3$ (iii) 2 cm^3

The Order of Operations (PEMDAS)

1) Solve what is within parentheses **(P)**.
2) Solve exponents **(E)**.
3) Solve multiplication **(M)** and division **(D)** from left to right.
4) Solve addition **(A)** and subtraction **(S)** from left to right.

You can remember PEMDAS with the silly mnemonic *Please Excuse My Dear Aunt Sally.*

Or make up your own!

5. Find the value of these expressions.

a. $120 - (9-4)^2$	c. $4 \cdot 5^2$	e. $10 \cdot 2^3 \cdot 5^2$
b. $120 - 9 - 4^2$	d. $(4 \cdot 5)^2$	f. $10 + 2^3 \cdot 5^2$
g. $(0.2 + 0.3)^2 \cdot (5-5)^4$	h. $0.7 \cdot (1-0.3)^2$	i. $20 + (2 \cdot 6 + 3)^2$

Example 1. Solve $(10 - (5-2))^2$.

First solve what is within the *inner* parentheses: $5 - 2 = 3$. We get $(10-3)^2$.

The rest is easy: $(10-3)^2 = 7^2 = 49$.

Example 2. Simplify $2 + \dfrac{1+5}{6^2}$.

Remember, the fraction line works like parentheses as a grouping symbol, grouping both what is above the line and also what is below it. First solve $1 + 5$, then the exponent.

$$2 + \frac{1+5}{6^2}$$

$$= 2 + \frac{6}{6^2} = 2 + \frac{1}{6} = 2\frac{1}{6}$$

6. Find the value of these expressions.

a. $(12 - (9-4)) \cdot 5$	c. $(10 - (8-5))^2$
b. $12 - (9 - (4+2))$	d. $3 \cdot (2 - (1-0.4))$

7. Find the value of these expressions.

a. $\dfrac{4 \cdot 5}{2} \cdot \dfrac{9}{3}$	b. $\dfrac{4 \cdot 5}{2} + \dfrac{9}{3}$	c. $\dfrac{4+5}{2} + \dfrac{9}{3-1}$

Expressions written using the ÷ symbol can be rewritten using the fraction line.
This usually makes them easier to read.

Example 3. In the expression $2 + 5 \cdot 2 \div 4 \cdot 10$, the division is by 4. This means that when written using the fraction line, only 4 goes in the denominator.

The expression becomes $2 + \dfrac{5 \cdot 2}{4} \cdot 10$

Here is how to simplify it:

$2 + \dfrac{5 \cdot 2}{4} \cdot 10$

$= 2 + \dfrac{10}{4} \cdot 10$

$= 2 + \dfrac{100}{4}$ (or $2 + 2.5 \cdot 10$)

$= 2 + 25 = 27$

Example 4. Rewrite the expression $2 \div 4 \cdot 3 \div (7 + 2) + 1$ using the fraction line.

Now there are *two* divisions: the first by 4 and the second by $(7 + 2)$. This means we will use <u>two</u> fractions in the expression.

It is written as $\dfrac{2}{4} \cdot \dfrac{3}{7+2} + 1$.

To simplify it, first calculate $7 + 2$, remembering that the fraction line implies parentheses around both the numerator and the denominator.

We get $\dfrac{2}{4} \cdot \dfrac{3}{9} + 1$.

Reducing the fractions, 2/4 equals 1/2, and 3/9 equals 1/3.

We get $\dfrac{1}{2} \cdot \dfrac{1}{3} + 1 = \dfrac{1}{6} + 1 = 1\dfrac{1}{6}$.

8. Rewrite each expression using a fraction line, then simplify. Compare the expression in the top row with the one below it. *Hint: Only what comes right after the "÷" sign goes into the denominator.*

a. $2 \div 5 \cdot 4$	b. $16 \div (2 + 6) \cdot 2$	c. $4 + 1 \div 3 + 2$
d. $2 \div (3 \cdot 4)$	e. $5 \div 9 \cdot 3 + 1$	f. $(1 + 3) \div (4 + 2)$

9. Find the value of these expressions. (Give your answer as a fraction or mixed number, not as a decimal.)

a. $\dfrac{9^2}{9} \cdot 6$	b. $\dfrac{2^3}{3^2}$	c. $\dfrac{(5-3) \cdot 2}{8 - 1 + 2} + 10$

10. Evaluate the expressions. (Give your answer as a fraction or mixed number, not as a decimal.)

a. $2x^2 - x$, when $x = 4$	**b.** $3s - 2t + 8$, when $s = 10$ and $t = 5$
c. $\dfrac{x^2}{x+1}$, when $x = 3$	**d.** $\dfrac{x+1}{x-1}$, when $x = 11$
e. $\dfrac{a+b}{b} + 2$, when $a = 1$ and $b = 3$	**f.** $\dfrac{n^2 + 2n}{n+3}$, when $n = 5$

11. Write a single mathematical expression ("number sentence") for each situation. Don't write just the answer.

a. You buy n hats for $4 each and m scarves for $6 each. Write an expression for the total cost. *cost =*	**b.** You have x pennies and y dimes in your pocket. What is their total value in cents? *value in cents =*
c. Molly and Mike share 10 cookies between them. Molly gets t cookies. Write an expression for how many cookies Mike gets. *Mike's cookies =*	**d.** Heather earns $11 per hour. Write an expression for how much she earns in n hours. *earnings =*
e. The club has 81 members, and 2/3 of them are girls. Write an expression for the number of girls. *girls =*	**f.** The club has n members, and 2/3 of them are girls. Write an expression for the number of girls. *girls =*
g. The price of a $60 book is discounted by 1/10. Write an expression for the current price. *price =*	**h.** The price of a book (cost x) is discounted by 1/10. Write an expression for the discounted price. *price =*
i. The altitude of a triangle is 3 and its base is b. Write an expression for its area (A). *A =*	**j.** The edge of a cube is c units long. Write an expression for its volume (V). *V =*

Expressions and Equations

Expressions in mathematics consist of: • numbers; • mathematical operations ($+$, $-$, $\cdot$, $\div$, exponents); • and letter variables, such as x, y, a, T, and so on. Note: Expressions do *not* have an "equals" sign! **Examples of expressions:** $\quad 5 \quad\quad \dfrac{xy^4}{2} \quad\quad T - 5$	An **equation** has two expressions separated by an equals sign: **(expression 1) = (expression 2)** **Examples:** $\quad 0 = 0 \quad\quad 2(a-6) = b$ $9 = -8$ $\quad\quad \dfrac{x+3}{2} = 1.5$ (a false equation)
What do we do with expressions? We can find the *value* of an expression (*evaluate* it). If the expression contains variables, we cannot find its value unless we know the value of the variables. For example, to find the value of the expression $2x$ when x is 3/7, we simply substitute 3/7 in place of x. We get $2x = 2 \cdot 3/7 = 6/7$. Note: When we write $2x = 2 \cdot 3/7 = 6/7$, the equals sign is *not* signaling an equation to solve. (In fact, we already know the value of x!) It is simply used to show that the value of the expression $2x$ here is the same as the value of $2 \cdot 3/7$, which is in turn the same as 6/7.	**What do we do with equations?** If the equation has a variable (or several) in it, we can try to *solve* the equation. This means we find the values of the variable(s) that make the equation true. For example, we can solve the equation $0.5 + x = 1.1$ for the unknown x. The value 0.6 makes the equation true: $0.5 + 0.6 = 1.1$. We say $x = 0.6$ is the **solution** or the **root** of the equation.

1. This is a review. Write an expression.

   a. $2x$ minus the sum of 40 and x.

   b. The quantity 3 times x, cubed.

   c. s decreased by 6

   d. five times b to the fifth power

   e. seven times the quantity x minus y

   f. the difference of t squared and s squared

   g. x less than 2 cubed

   h. the quotient of 5 and y squared

   i. 2 less than x to the fifth power

   j. x cubed times y squared

   k. the quantity $2x$ plus 1 to the fourth power

   l. the quantity x minus y divided by the quantity x squared plus one

> To read the expression $2(x + y)$, use the word ***quantity***:
> "two times the quantity x plus y."
>
> There are other ways, as well, just not as common:
>
> "two times the sum of x and y," or
> "the product of 2 and the sum x plus y."

Some equations are *true*, and others are *false*. For example, 0 = 9 is a false equation.

Some equations are neither. The equation $x + 1 = 7$ is neither false nor true in itself. However, if x has a specific value, then we can tell if the equation becomes true or false.

Indeed, solving an equation means finding the values of the variables that make the equation *true*. The solutions of the equation can also be called its **roots**.

Example. Find the root of the equation $20 - 2y^2 = 2$ in the set {0, 1, 2, 3, 4}.

Try each number from the set, checking to see if it makes the equation true:

$20 - 2 \cdot 0^2 = 20 \neq 2$ $20 - 2 \cdot 2^2 = 12 \neq 2$ $20 - 2 \cdot 4^2 = -12 \neq 2$

$20 - 2 \cdot 1^2 = 18 \neq 2$ $20 - 2 \cdot 3^2 = 2$

So, in the given set, the only root of the equation is 3.

2. Write an equation. Then solve it.

	Equation	Solution
a. 78 decreased by some number is 8.		
b. The difference of a number and 2/3 is 1/4.		
c. A number divided by 7 equals 3/21.		

3. **a.** Find the root(s) of the equation $n^2 - 9n + 14 = 0$ in the set on the right.

    1   10   3
     6   2   7

   b. Find the root(s) of the equation $9x - 5 = 2x$ in the set on the right.

    1/5   5/7
      7/9

4. Which of the numbers 0, 1, 3/2, 2 or 5/2 make the equation $\dfrac{y}{y-1} = 3$ true?

5. **a.** Ann is 5 years older than Tess, and Tess is n years old. Write an <u>expression</u> for Ann's age.

   b. Let A be Alice's age and B be Betty's age. Find the <u>equation</u> that matches the sentence "Alice is 8 years younger than Betty."

   $A = 8 - B$ $A = B - 8$ $B = A - 8$

   Hint: give the variables some test values.

6. **a.** In a bag of blocks, there are twice as many red blocks as there are blue blocks, and three times as many green blocks as blue blocks. Let's denote the number of blue blocks with x. Write an expression for the amount of red blocks.

b. Write another expression for the amount of green blocks.

7. **a.** Timothy earns s dollars in a month. He pays 1/5 of it in taxes and gets to keep 4/5 of it. Write an expression for the amount Timothy gets to keep.

b. Write another, different expression for the amount Timothy gets to keep.
(Hint: if you used a fraction in a., use a decimal now, or vice versa).

8. Circle the equation that matches the situation. *Hint: give the variable(s) some value(s) to test the situation.*

a. The price of a phone is discounted by 1/4, and now it costs $57.

| $\frac{p}{4} = \$57$ | $\frac{3p}{4} = \$57$ | $\frac{4p}{3} = \$57$ | $\frac{p}{3} = 4 \cdot \$57$ | $p - 14 = \$57$ |

b. Matt bought three computer mice for $25 each and five styluses for p dollars each. He paid a total of $98.

| $25 + 5p = 98$ | $3 \cdot 25 + 5p = 98$ | $3p + 125 = 98$ |
| $3 \cdot 25 \cdot 5p = 98$ | $3 \cdot 25 + p = 98$ | $5p + 75 = 98$ |

c. Jeremy sells fresh-squeezed orange juice for x per glass. Today he has discounted the glass of juice by $1. A customer buys three glasses, and the total comes to $5.40.

| $3(x - \$1) = \$5.40$ | $3x - \$1 = \$5.40$ |
| $x - \$1 = 3 \cdot \$5.40$ | $3(x - 0.1) = \$5.40$ |

Here is a very strange equation: $n = n$
If you think about it, you can put *any* number in place of n, and the equation will be true!

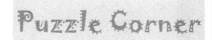

For example, if n is 5, we get 5 = 5 (a true equation). This equation has an **infinite number of solutions**—any number n will make it true!

Find the equations below that also have an infinite number of solutions.

$1 + x = 2 + x$ $4 + c + 1 = 2 + 3 + c$ $2y - 10 = y + y$

$3z - 1 = z - 1 + z + z$ $6 + 2n + 3n = n + 6$ $b \cdot b = 0$

Properties of the Four Operations

In this lesson, we will look at some special properties of the basic operations. You already know them. In fact, you have been using them since you first learned to add! But this time, we will name the properties and study them in detail.

1. Addition is **commutative**.

If your father commutes to work, he changes where he is at from home to work and back. In math, when a and b commute, they change places. This means that you can change the order of the addends when you add two numbers. In symbols: $a + b = b + a$.

In other words: When adding 2 numbers, you can change their order.

2. Addition is **associative**.

When you associate with people, you group yourself with them. In math, when a and b associate, they are grouped together. The associative property says that when adding three numbers, it does not matter if you begin by adding the first two or the last two. In symbols: $(a + b) + c = a + (b + c)$.

(What about adding a and c first? Would that work?)

Then we have the identical properties for multiplication.

3. Multiplication is **commutative**.

When multiplying two numbers, you can change their order. In symbols: $ab = ba$.

4. Multiplication is **associative**.

When multiplying three numbers, it does not matter if you start with the first two or if you start with the last two. In symbols: $(ab)c = a(bc)$.

(Could you even start by multiplying a and c first? Would that work?)

1. Are the two expressions in each box equivalent? That is, do they have the same value for any value of c? Give c some test values to check.

a. $c + 5$	b. $c - 5$	c. $c \div 6$	d. $5c$
$5 + c$	$5 - c$	$6 \div c$	$c \cdot 5$

2. Is subtraction commutative? In other words, will $a - b$ always have the same value as $b - a$, no matter what values we give to a and b? Explain your reasoning.

3. Is division commutative? Does $a \div b$ always have the same value as $b \div a$ for any numbers that we might use for a and b? Explain your reasoning.

4. Are the expressions equal, no matter what value *n* has? Give *n* some test values to check.

a. $(n + 2) + 5$	b. $(n \cdot 2) \cdot 5$	c. $(n - 7) - 3$
$n + (2 + 5)$	$n \cdot (2 \cdot 5)$	$n - (7 - 3)$

5. **a.** Name the property of arithmetic illustrated by (a) above.

   b. Name the property of arithmetic illustrated by (b) above.

6. Are the expressions equal, no matter what values the variable(s) have? If so, you don't need to do anything else. If not, provide a counterexample: specific values of the variables that show the expressions do NOT have the same value.

a. $n - 2 - m$	b. $(2n + 1) \cdot 5$
$\quad n + (2 - m)$	$\quad 5 \cdot (1 + 2n)$
Not equal. For example when $n = 5$ and $m = 1$, we get: $n - 2 - m = 5 - 2 - 1 = \underline{2}$, but $n + (2 - m) = 5 + (2 - 1) = \underline{6}$.	
c. $(n - 2) \cdot m$	d. $a + 2b$
$\quad m(2 - n)$	$\quad b + 2a$

7. Is subtraction associative? In other words, is it true that $(a - b) - c$ has the same value as $a - (b - c)$, no matter what values *a*, *b*, and *c* get? Explain your reasoning.

8. Is division associative? In other words, is it true that $(a \div b) \div c$ has the same value as $a \div (b \div c)$, no matter what values *a*, *b*, and *c* get? Explain your reasoning.

Since addition is both commutative and associative, it follows that **we can add a list of numbers in any order** we choose. Of course, you already knew that!

(Optional) Here is a proof that, when you add three numbers, you can start with the first and last numbers.

Consider the sum of three numbers $a + b + c$.

1. Because addition is <u>commutative</u>, we can switch the order of a and b.

   Thus $b + a + c$ has the same value as $a + b + c$.

2. Because addition is associative, $b + a + c = b + (a + c)$. So instead of proceeding from left to right and adding the first two numbers first, we can add the last two numbers first.

3. Therefore we can indeed calculate the sum $a + c$ first because it is inside the parentheses.

The same is true of multiplication: **you can multiply a list of numbers in any order you choose.**

We can use these properties of operations to simplify expressions.

Example 1. Simplify $5 + a + b + 7 + a$.

Here is a model for this expression: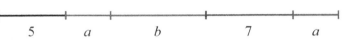

Because we can add in any order, we can add $5 + 7$ to get 12. Moreover, we can add $a + a$ and write that as $2a$. So $5 + a + b + 7 + a$ simplifies to $2a + b + 12$. That is as simplified as it can get.

Note: It's customary to write the terms with variables in alphabetical order and put the constant term (here, the "12") last.

9. Write an expression from the illustration and simplify it.

a.	$x \quad x \quad x \quad x \quad x$
b.	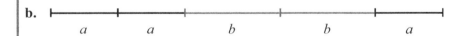
c.	$s \quad s \quad 15 \quad s \quad s$
d.	

10. Simplify the expressions.

a. $5 + v + 8 + v + v$	**b.** $e + e + 9 + e + 28 + e$	**c.** $2v + v$
d. $5a + 8a$	**e.** $8 + 6a + 5b + 3b + 9a$	**f.** $10t + s + 2 + s + 3s$

Example 2. Simplify $x \cdot x \cdot 4 \cdot y \cdot y \cdot y \cdot y$.

The part $x \cdot x$ can be written as x^2, and the part $y \cdot y \cdot y \cdot y$ as y^4.
Those are multiplied, so in total we get $x^2 \cdot 4 \cdot y^4$.

Since the multiplication sign is usually omitted from between variables, and the number you multiply by (4) is usually written in front of the term, we would write this as $4x^2y^4$.

11. Simplify the expressions.

a. $a \cdot a \cdot 7$	**b.** $2 \cdot s \cdot s \cdot 8$	**c.** $a \cdot a \cdot d \cdot d \cdot d \cdot d$

Error alert!

Multitudes of algebra students have confused $b \cdot b \cdot b$ with $b + b + b$ and written $b \cdot b \cdot b = 3b$.

That is NOT true! Make sure you understand and remember the reason why:

Just as $b + b + b$ is repeated addition, for which the shortcut is to multiply: $b + b + b = 3b$; so, too, $b \cdot b \cdot b$ is repeated multiplication, for which the shortcut is to use an exponent: $b \cdot b \cdot b = b^3$.

d. $y \cdot x \cdot x \cdot y \cdot 2 \cdot y \cdot x$	**e.** $d + a + a + d$	**f.** $z + z + z + 8 + z$
g. $y \cdot y \cdot y \cdot 8 \cdot t \cdot t$	**h.** $b \cdot b \cdot 9 \cdot b \cdot 3 \cdot 1 \cdot b$	**i.** $2s + s + t + 3s + 2$

12. Are the expressions equal, no matter what values x and y have? If yes, you don't need to do anything else. If not, provide a counterexample.

a. $\dfrac{5}{x}$ $\dfrac{x}{5}$	**b.** $x + \dfrac{y}{2}$ $y + \dfrac{x}{2}$
c. $\dfrac{x+y}{2}$ $\dfrac{x}{2} + \dfrac{y}{2}$	**d.** $\dfrac{x}{x}$ $\dfrac{y}{y}$

13. Summary. Write "yes" or "no" to indicate if the operation is commutative or associative. Include examples or comments if you want to.

Operation	Commutative?	Associative?	Optional notes/examples
addition			
subtraction			
multiplication			
division			

Simplifying Expressions

Example. Simplify $2x \cdot 4 \cdot 5x$.

Notice, this expression contains only multiplications (because $2x$ and $5x$ are also multiplications).

Since we can multiply in any order, we can write this expression as $2 \cdot 4 \cdot 5 \cdot x \cdot x$.

Now we multiply 2, 4, and 5 to get 40. What is left to do? The part $x \cdot x$, which is written as x^2.

So, $2x \cdot 4 \cdot 5x = 40x^2$.

Note: The equals sign used in $2x \cdot 4 \cdot 5x = 40x^2$ signifies that the two expressions are equal no matter what value x has. That equals sign does not signify an equation that needs to be solved.

Similarly, we can simplify the expression $x + x$ and write $2x$ instead. That whole process is usually written as $x + x = 2x$.

Again, the equals sign there does not indicate an equation to solve, but just the fact that the two expressions are equal. In fact, if you think of it as an equation, *any* number x satisfies it! (Try it!)

1. Simplify the expressions.

a. $p + 8 + p + p$	**b.** $p \cdot 8 \cdot p \cdot p \cdot p$	**c.** $2p + 4p$
d. $2p \cdot 4p$	**e.** $5x \cdot 2x \cdot x$	**f.** $y \cdot 2y \cdot 3 \cdot 2y \cdot y$

2. Write an expression for the area and perimeter of each rectangle.

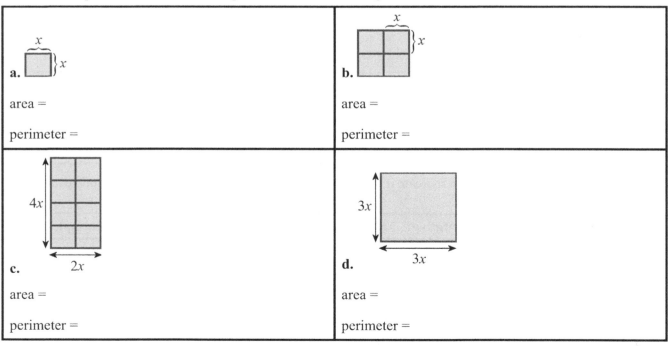

22

3. **a.** Sketch a rectangle with sides $2b$ and $7b$ long.

   b. What is its area?

   c. What is its perimeter?

4. **a.** The perimeter of a rectangle is $24s$.
   Sketch one such rectangle.

   What is its area?
   Hint: there are many possible answers.

   b. Find the area and perimeter of your rectangle in (a) if s has the value 3 cm.

5. **a.** Which expression below is for an area of a rectangle? Which one is for a perimeter?

   $4a + 4b$ $2a \cdot 2b$

   b. Sketch the rectangle.

6. **a.** Find the value of the expressions $3p$ and $p + 3$ for different values of p.

Value of p	$3p$	$p + 3$
0		
0.5		
1		
1.5		
2		
2.5		
3		
3.5		
4		

   b. Now, look at the table. Can you tell which is larger, $3p$ or $p + 3$?

Some review! In algebra, a **term** is an expression that consists of numbers and/or variables that are multiplied together. A single number or variable is also a term.

Examples.
- $2xy$ is a term, because it only contains multiplications, a number, and variables.
- $(5/7)z^3$ is a term. Remember, the exponent is a shorthand for repeated multiplication.
- Addition and subtraction separate the individual terms in an expression from each other. For example, the expression $2x^2 - 6y^3 + 7xy + 15$ has four terms, separated by the plus and minus signs.
- $s + t$ is *not* a term, because it contains addition. Instead, it is a sum of *two* terms, s and t.

The number by which a variable or variables are multiplied is called a ***coefficient***.

Examples.
- The term $0.9ab$ has the coefficient 0.9.
- The coefficient of the term m^2 is 1, because you can write m^2 as $1 \cdot m^2$.

If the term is a single number, such as 7/8, we call it a ***constant***.

Example. The expression $1.5a + b^2 + 6/7$ has three terms: $1.5a$, b^2, and $6/7$. The last term, $6/7$, is a constant.

7. Fill in the table.

Expression	the terms in it	coefficient(s)	Constants
$(5/6)s$			
w^3			
$0.6x + 2.4y$			
$x + 3y + 7$			
$p \cdot 101$			
$x^5y^2 + 8$			

The two terms in the expression $2x^2 + 5x^2$ are **like terms**: they have the same variable part (x^2). Because of that, we can add the two terms to simplify the expression. To do that, simply add the coefficients 2 and 5 and use the same variable part: $2x^2 + 5x^2 = 7x^2$. It is like adding 2 apples and 5 apples.

However, you cannot add (or simplify) $2x + 7y$. That would be like adding 2 apples and 7 oranges.

Example. Simplify $6x - x - 2x + 9x$. The terms are like terms, so we simply add or subtract the coefficients: $6 - 1 - 2 + 9 = 12$ and tag the variable part x to it. The expression simplifies to $12x$.

8. Simplify the expressions.

a. $5p + 8p - p$	**b.** $p^2 + 8p^2 + 3p^2$	**c.** $12a^2 - 8a^2 - 3a^2$

9. Write an expression for the total area.

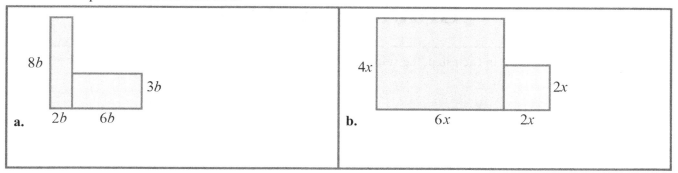

In the following problems, write an expression for part (a), and then for part (b) write an equation and solve it. Don't skip writing the equation, even if you can solve the problem without it, because we are practicing writing equations! You don't have to use algebra to solve the equations—you can solve them in your head or by guessing and checking.

10. **a.** The length of a rectangle is 4 meters and its width is w. What is its perimeter? Write an expression.

    b. Let's say the perimeter has to be 22 meters. How wide is the rectangle then? Write an *equation* for this situation, using your expression from (a).

    Remember, you do not have to use algebra to solve the equation—you can solve it in your head or by "guess and check." But do write the equation.

11. **a.** Linda borrows six books from the library each week, and her mom borrows two. How many books, in total, do both of them borrow in w weeks? Write an expression.

    b. How many weeks will it take them to have borrowed 216 books? Write an equation.

12. **a.** Alice buys y containers of mints for $6 apiece. A fixed shipping cost of $5 is added to her order. What is the total cost? Write an expression.

    b. The total bill was $155. How many containers of mints did she buy? Write an equation.

Puzzle Corner

a. What is the total value, in cents, if Ashley has n dimes and m quarters? Write an expression.

b. The total value of Ashley's coins is 495 cents. How many dimes and quarters can she have?
   Hint: make a table to organize the possibilities.

Growing Patterns 1

This is a pattern of squares. In each step, the side of the square grows by 1 flower.

Step   1   2   3   4   5

Draw steps four and five.

How many flowers will there be in step 13?

How many flowers will there be in step n?
(If you get stuck, the answer is at the bottom of the page, but try it on your own first.)

Here is another pattern:

Step   1   2   3   4   5

Draw steps four and five.

How do you see this pattern growing? (There's more than one way to look at it!)

How many flowers will there be in step 39?

What about in step n?

(The answers are in the next box, but stop here to try the problem yourself first.)

---

1. One way is to see the pattern as an L-shape. This means in step 39 there are 39 flowers horizontally and 39 vertically, but with 1 flower overlapping, so the overlapping flower needs to be subtracted. In total, there are $39 + 39 - 1 = 77$ flowers in step 39.

   Similarly, in step n, there are $n + n - 1$ flowers, or $2n - 1$ flowers.

2. Another way to see it is as a square from which is subtracted a smaller square. This means that, in step 39, we have a square that is $39 \cdot 39$, and then a square $38 \cdot 38$ is taken away. $39 \cdot 39 - 38 \cdot 38 = 1521 - 1444 = 77$.

   In step n the length of the side of the square is n, so its area is n^2. The square that is "taken away" has a side $n - 1$ long (one less than the side of the bigger square). Its area is $(n - 1)^2$.

   Then we subtract these two areas to find that there are $n^2 - (n - 1)^2$ flowers in each step.

   Of course, this expression is more complicated than $2n + 1$, but believe it or not, the two *are* equivalent!

Answer for the pattern at the top of this page: There are n^2 flowers in step n.

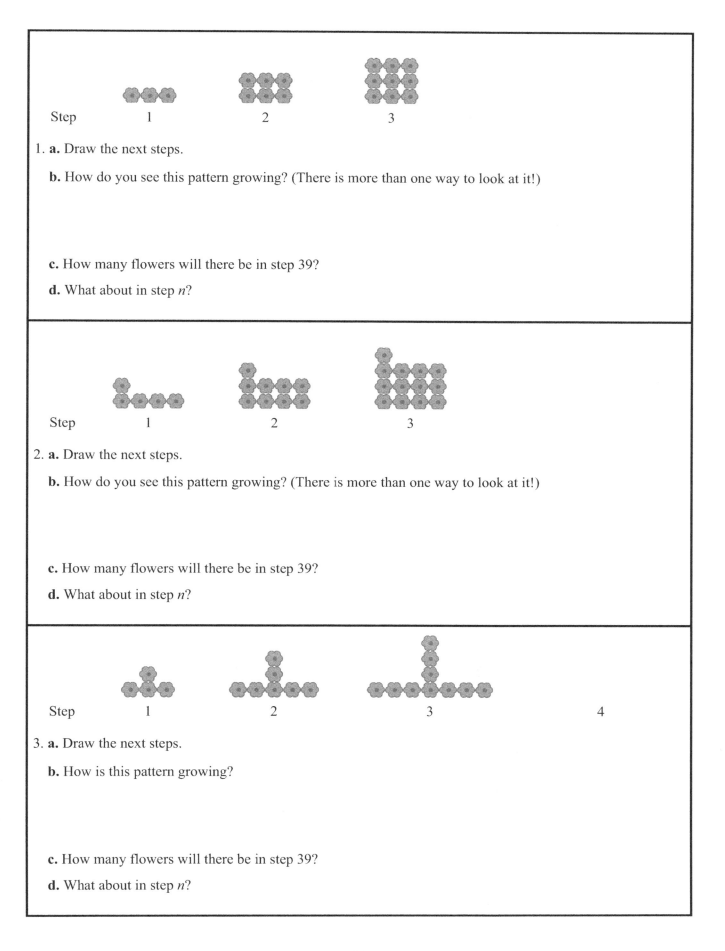

Step 1 2 3

1. **a.** Draw the next steps.

   b. How do you see this pattern growing? (There is more than one way to look at it!)

   c. How many flowers will there be in step 39?

   d. What about in step *n*?

Step 1 2 3

2. **a.** Draw the next steps.

   b. How do you see this pattern growing? (There is more than one way to look at it!)

   c. How many flowers will there be in step 39?

   d. What about in step *n*?

Step 1 2 3 4

3. **a.** Draw the next steps.

   b. How is this pattern growing?

   c. How many flowers will there be in step 39?

   d. What about in step *n*?

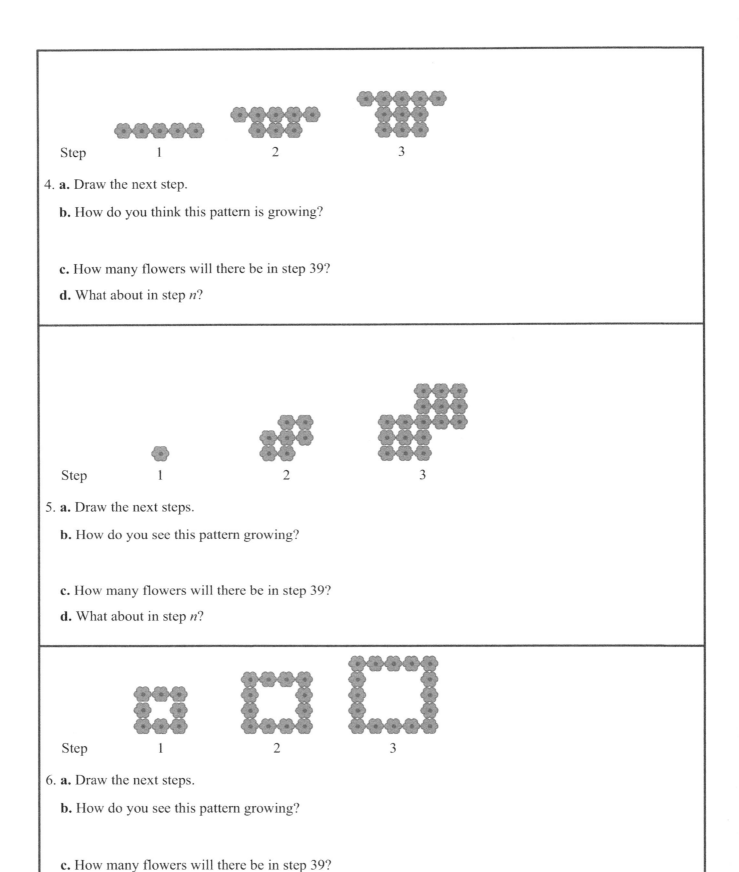

Step   1   2   3

4. **a.** Draw the next step.

   b. How do you think this pattern is growing?

   c. How many flowers will there be in step 39?

   d. What about in step n?

Step   1   2   3

5. **a.** Draw the next steps.

   b. How do you see this pattern growing?

   c. How many flowers will there be in step 39?

   d. What about in step n?

Step   1   2   3

6. **a.** Draw the next steps.

   b. How do you see this pattern growing?

   c. How many flowers will there be in step 39?

   d. What about in step n?

The Distributive Property

The **distributive property** states that $a(b + c) = ab + ac$ for any numbers a, b, and c.

It says we can *distribute* multiplication over addition. This means that instead of multiplying a times the sum $b + c$, we can multiply the numbers b and c separately by a and add last.

Example 1. The expression $20(x + 5)$ is equal to $20x + 20 \cdot 5$, which simplifies to $20x + 100$.

Notice what happens: Each term in the sum $(x + 5)$ gets multiplied by the factor 20! Graphically:

$$20(x + 5) = 20x + 20 \cdot 5$$

Example 2. To multiply $2a(3 + c)$ using the distributive property, we need to multiply **both** 3 and c by $2a$:

$$2a(3 + c) = 2a \cdot 3 + 2a \cdot c$$

Lastly, we simplify: $2a \cdot 3$ simplifies to $6a$, and of course we can write $2a \cdot c$ without the multiplication sign:

$$2a \cdot 3 + 2a \cdot c = 6a + 2ac$$

Here is a way to **model the distributive property using line segments.**

The model shows a pattern of line segments of lengths x and 1 repeated four times.
In symbols, we write $4 \cdot (x + 1)$.

However, it is easy to see that the total length can *also* be written as $4x + 4$.

Therefore, $4 \cdot (x + 1) = 4x + 4$.

1. Write an expression for the repeated pattern in the model. Then multiply the expression using the distributive property.

a.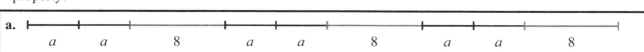

$3(2a + 8) =$

b.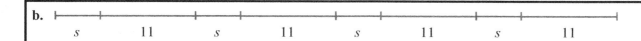

c.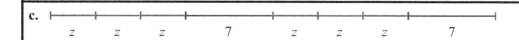

2. Draw line segments to represent the expressions. Then, multiply the expressions using the distributive property.

a. $3(b + 8) =$
b. $4(2w + 1) =$
c. $2(3x + 5) =$

3. Use the distributive property to multiply.

a. $2(x + 9) =$	**b.** $7(4y + 5) =$	**c.** $10(9x + 8) =$
d. $8(x + y) =$	**e.** $s(4 + t) =$	**f.** $u(v + w) =$

4. The side of a regular hexagon is $2x + 4$. What is its perimeter?

5. The perimeter of a square is $24y + 40$. How long is its side?

$P = 24y + 40$

6. **a.** Write an expression for the total cost of buying n jars of coconut oil for $20 each.

   b. What is the total cost, if an additional shipping cost of $11 is added to the order?

   c. You repeat the same order three times during the year.
   Multiply the expression from (b) by 3, and use the distributive property.

   d. How many jars of coconut oil did you buy in a year, if you spent $393 in these three orders?

The distributive property works the same way *with subtraction* and with *more than two terms* in the second factor. The proof that it works with subtraction has to do with how negative numbers work in multiplication, and it is not presented here.

However, consider this numerical example. To multiply 7 · 98, think of the 98 as 100 − 2, and multiply in parts:
$$7(100 - 2) = 7 \cdot 100 - 7 \cdot 2 = 700 - 14 = 686$$

Example 3. $2(7x - y) = 2 \cdot 7x - 2 \cdot y = 14x - 2y$

Example 4. $8(s - t + 5) = 8s - 8t + 40$

7. Use the distributive property to multiply.

a. $11(x - 7) =$	**b.** $30(x + y + 5) =$	**c.** $10(r + 2s + 0.1) =$
d. $5(3x - 2y - 6) =$	**e.** $s(1.5 + t - x) =$	**f.** $0.5(3v + 2w - 7) =$

8. Solve mentally! (Hint: the distributive property will help.)

a. $8 \cdot 99 =$	**b.** $6 \cdot 98 =$	**c.** $5 \cdot 599 =$

The area of this *whole* rectangle is 9 times $(8 + b)$. But, if we think of it as *two* rectangles, the area of the first rectangle is $9 \cdot 8$, and of the second, $9 \cdot b$.

Of course, these two expressions are equal:
$$9 \cdot (8 + b) = 9 \cdot 8 + 9 \cdot b = 72 + 9b \text{ or } 9b + 72.$$

9. Write an expression for the area in two ways, thinking of the overall rectangle or its component rectangles.

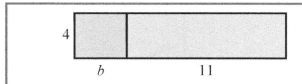

a. ____(_____ + _____) and

4 · _____ + 4 · _____ =

b. ____(_____ + _____) and

____ · _____ + ____ · _____ =

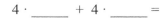

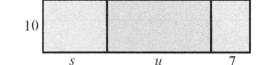

c. ____(_____ + _____ + _____) and

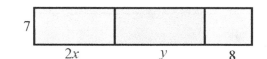

d. and

10. Find the missing numbers or variables in these area models.

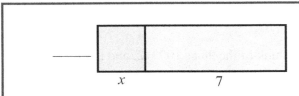

a. ____ (x + 7) = ____ x + 63

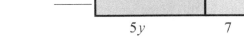

b. ____ (5y + 7) = 40y + _____

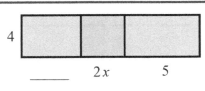

c. The total area is 12y + 8x + 20.

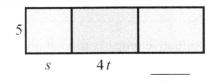

d. 5(s + 4t + ____) = 5s + 20t + 15v

11. Find the missing number or term in the equations.

a. ____ (20x + 3) = 200x + 30

b. 4(6s − ____) = 24s − 4x

c. 2(____ + 1.5y + 0.9) = 7x + 3y + 1.8

d. 4(____ − ____ + ____) = 0.4x − 1.2y + 1.6

12. Use the distributive property "backwards" to write the expression as a product. This is called **factoring**.

a. 2x + 6 = ____ (x + 3)

b. 4y + 16 = 4(____ + ____)

c. 21t + 7 = ____ (3t + ____)

d. 16d + 24 = ____ (2d + ____)

e. 15x − 35 = ____ (____ − ____)

f. 7a − 49 =

13. **a.** Sketch a rectangle with an area of 9x + 15.

b. Sketch a rectangle with an area of 9a + 15b + 3.

14. Factor these sums (write them as products). Think of divisibility!

a. 64x + 40 =

b. 54x + 18 =

c. 100y − 20 =

d. 90t + 33s + 30 =

e. 36x − 12y + 24 =

f. 2x + 8z − 40 =

The distributive property works with division, too. Just like we can multiply in parts, we can *divide in parts*.

Example 5. $\dfrac{50+35}{5}$ is the same as $\dfrac{50}{5} + \dfrac{35}{5}$. You can then do the two divisions separately and add last.

Example 6. $\dfrac{6x+y}{2}$ is the same as $\dfrac{6x}{2} + \dfrac{y}{2}$. And $\dfrac{6x}{2}$ simplifies to $3x$.

This works because any division can be rewritten as a multiplication by a fraction, and multiplication is distributive.

15. Divide in parts using mental math. You may end up with a fraction in the answer.

a. $\dfrac{300+2}{3}$	b. $\dfrac{13+700}{7}$	c. $\dfrac{5{,}031}{5}$
d. $\dfrac{5x-3}{6}$	e. $\dfrac{x+7}{7}$	f. $\dfrac{4x+2}{4}$

16. The Larson family are planning their new house. It is going to be 25 ft on one side and have a garage that is 15 ft wide, but they have not decided on the length of the house yet.

    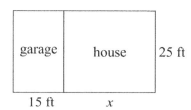

    a. If the total area of the house + garage is limited to 1200 square feet, how long can the house be?

    b. Write a single equation for the question above. Write it in the form "(formula for area) = 1200." You do not have to solve the equation—just write it.

We can even model expressions with subtraction, such as $3(7-2)$, using an area model. We use dark shading to indicate that an area is subtracted ("negative" area).

Puzzle Corner

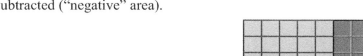

For example, the picture above illustrates $3(7-2) = 3 \cdot 7 - 3 \cdot 2$.

a. What expression is modeled below? 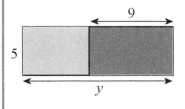	b. Draw a model for $3(x + y - 2) = 3x + 3y - 6$.

Chapter 1 Review

1. Find the value of these expressions.

a. $(6+4)^2 \cdot (12-9)^3$	b. $3 \cdot (5-(7-5))$	c. $\dfrac{(5-3) \cdot 2}{2^3} + 7$

2. Name the property of arithmetic illustrated by the fact that $(5 \cdot z) \cdot 3$ is equal to $5 \cdot (z \cdot 3)$.

3. Evaluate the expressions.

a. $100 - 2x^2$, when $x = 5$	b. $\dfrac{2s}{s^3 + 3}$, when $s = 4$

4. Which equation matches the situation? *Hint: give the variable(s) some value(s) to test the situation.*

   a. The shorter beam (length l_1) is 1.5 meters shorter than the longer beam (length l_2).

$l_1 = 1.5 - l_2$	$l_2 = 1.5 - l_1$	$l_2 = l_1 - 1.5$	$l_1 = l_2 - 1.5$

   b. The dog lost 1/6 of its original weight (w), and weighs now 23 kg.

$\dfrac{w}{6} = 23$	$\dfrac{5w}{6} = 23$	$\dfrac{6w}{5} = 23$	$w - 1/6 = 23$	$w - 5/6 = 23$

5. Is subtraction commutative? In other words, is it true that $a - b$ has the same value as $b - a$, no matter what values we use for a and b? Explain your reasoning.

6. Write a SINGLE expression to match these situations.

   a. A pair of jeans costs p dollars. The jeans are now discounted by 1/5 of that price. Write an expression for the discounted price.

   b. It costs Mandy $0.18 to drive her car one mile. How much does it cost her to drive y miles? Write an expression.

   c. The pet store sells dog food in bags of two different sizes: 3-kg and 8-kg. What is the total weight of n of the smaller bags and m of the larger bags?

7. Simplify the expressions.

a. $x + 2 + x + x$	b. $x \cdot 2 \cdot x \cdot x \cdot x$	c. $8v + 12v$
d. $8v \cdot 12v$	e. $4z \cdot 9z \cdot z$	f. $f + 2f + 10g - f - 4g$

8. **a.** Sketch a rectangle that is $5x$ tall and $2x$ wide.

   b. What is its area?

   c. What is its perimeter?

9. Use the distributive property to multiply.

a. $12(v - 9)$	b. $3(a + b + 2)$	c. $3(0.5t - x)$

10. Draw a diagram of two rectangles to illustrate that the product $11(x + 7)$ is equal to $11x + 77$.

11. Fill in the table.

Expression	the terms in it	coefficient(s)	Constants
a^8			
$2x + 9y$			

12. The perimeter of a regular pentagon is $30s + 45$. How long is one side?

13. Factor these sums (write them as products). Think of divisibility!

a. $48x + 12 =$	b. $40x - 25 =$
c. $6y - 2z =$	d. $56t - 16s + 8 =$

Chapter 2: Integers
Introduction

This chapter deals with integers, which are signed (positive or negative) whole numbers. We begin with a review of addition and subtraction of integers from 6th grade. Then we study in detail multiplication and division of integers and conclude with negative fractions and the order of operations.

The first lesson reviews the concepts of integers, absolute value, the opposite of an integer, and simple inequalities on a number line. The next lessons present the addition and subtraction of integers through two visual models: first as movements on a number line, and then using positive and negative counters. These lessons also endeavor to connect the addition and subtraction of integers with various situations from real life.

The lesson *Subtraction of Integers* includes this important principle: Any subtraction can be converted into an addition (of the number with the opposite sign) and vice versa. This principle allows us to calculate not only subtractions such as $5 - (-7)$ but also problems that contain both addition and subtraction. These mixed problems become simple sums after the subtractions have been converted into additions. Converting subtractions into additions or vice versa is also important when simplifying expressions. For example, $5 + (-x)$ can be simplified to $5 - x$.

Next, we study the distance between two integers. This can be found by taking the absolute value of their difference: the distance between x and y is $|x - y|$. Students learn to use this formula to find distances between integers, and they also compare the result the formula gives to the answer they get by logical thinking.

The lesson *Multiplying Integers* not only teaches the mechanics of how to multiply integers but also gives both intuitive understanding and formal justification for the shortcut, "a negative times a negative makes a positive." This formal justification using the distributive property introduces and illustrates the type of careful and precise reasoning that mathematicians use in proofs.

Then, the lesson on the division of integers leads naturally into the topic of negative fractions. The final topic is a review of the order of operations: we perform several operations at a time with integers.

You can find matching videos for the topics in this chapter at
https://www.mathmammoth.com/videos/prealgebra/pre-algebra-videos.php#integers

The Lessons in Chapter 2

	page	span
Integers	41	*4 pages*
Addition and Subtraction on the Number Line	45	*4 pages*
Addition of Integers	49	*3 pages*
Subtraction of Integers	52	*4 pages*
Adding and Subtracting Many Integers	56	*2 pages*
Distance and More Practice	58	*4 pages*
Multiplying Integers	62	*5 pages*
Dividing Integers	67	*2 pages*
Negative Fractions	69	*4 pages*
The Order of Operations	73	*2 pages*
Chapter 2 Mixed Review	75	*2 pages*
Chapter 2 Review	77	*3 pages*

Helpful Resources on the Internet

You can also access this list of links at **https://links.mathmammoth.com/gr7ch2**

Free Downloadable Integer Fact Sheets
https://www.homeschoolmath.net/download/Add_Subtract_Integers_Fact_Sheet.pdf

https://www.homeschoolmath.net/download/Multiply_Divide_Integers_Fact_Sheet.pdf

ORDERING INTEGERS

Number Climb
Click the balls in the ascending order of numbers.
https://www.mathplayground.com/mobile/numberballs_fullscreen.htm

Plot Inequalities Quiz
Practice plotting simple inequalities, such as $x < -7$, on a number line in this 10-question interactive quiz.
https://www.thatquiz.org/tq-o/?-j18-l3-p0

Compare Integers Quiz
A 10-question online quiz where you compare two integers. You can also modify the quiz parameters to include sums, differences, products, and quotients, which makes it more challenging.
http://www.thatquiz.org/tq-8/?-j11-l1i-p0

ABSOLUTE VALUE AND OPPOSITES

Number Opposites Challenge
Practice solving challenging problems finding the opposites of numbers.
https://bit.ly/Number-Opposites-Challenge

Absolute Value to Find Distance
Practice taking the absolute value of the difference of two numbers to find the distance between those numbers. Apply this principle to solve word problems.
https://www.khanacademy.org/math/arithmetic-home/negative-numbers/abs-value/e/absolute-value-to-find-distance

Absolute Value Quiz
Find the absolute value of each integer or sum.
https://www.softschools.com/quizzes/math/absolute_value/quiz1035.html

Absolute Value Exercises
Click on "new problem" to get randomly generated practice-problems that practice absolute value.
https://www.onemathematicalcat.org/algebra_book/online_problems/intro_abs_val.htm#exercises

Absolute Value Quiz - harder
This quiz includes mixed operations and absolute value.
https://www.softschools.com/testing/math/theme2.html

ADDITION AND SUBTRACTION

Interactive Number-Line
This app helps students visualize number sequences and illustrate strategies for counting, comparing, adding, subtracting, multiplying, and dividing. Choose number lines labeled with whole numbers, fractions, decimals, or negative numbers.
https://apps.mathlearningcenter.org/number-line/

Number-Line Addition
Click on the addition sentence on the fruit that matches the number line. Choose level 3.
https://www.sheppardsoftware.com/math/integers/fruit-splat-numberline-addition/

Number-Line Integer Subtraction
Click on the subtraction sentence on the fruit that matches the number line. Choose level 3.
https://www.sheppardsoftware.com/math/integers/fruit-splat-numberline-subtraction/

Add Negative Numbers on a Number Line (from Khan Academy)
Practice matching number line diagrams to addition expressions involving negative numbers.
https://bit.ly/add-negative-numbers-number-line

Fruit Splat Integers Addition
Practice adding integers with this interactive online game.
https://www.sheppardsoftware.com/math/integers/fruit-splat-addition/

Spider Match
Choose pairs of numbers that add to the given integer. Can be played as a multi-player game or against the computer.
https://www.arcademics.com/games/spider-match

Interpreting Negative Number Statements (from Khan Academy)
Practice matching addition and subtraction equations to real-world scenarios in this interactive exercise.
https://bit.ly/interpret-negative-number-statements

Add Integers Quiz
Practice adding integers in this 10-question online quiz.
https://www.thatquiz.org/tq-1/?-j4101-lk-p0

Understanding Subtraction as Adding the Opposite
Practice your integer skills with this interactive quiz.
https://bit.ly/Understanding-Subtraction-as-Adding-the-Opposite

Subtracting Negative Integers
Practice subtracting positive and negative single-digit numbers in this interactive online activity.
https://bit.ly/Subtracting-Negative-Integers

Adding & Subtracting Several Negative Numbers
Practice solving addition and subtraction problems with several integers.
https://bit.ly/Adding-Subtracting-Several-Negative-Numbers

Find the missing integers
Fill in the missing integer in addition equations such as $-23 + \underline{} = -8$.
https://www.aaamath.com/g8_65_x3.htm

Find the Missing Number (from Khan Academy)
Practice finding the missing value in an addition or subtraction equation involving negative numbers.
https://bit.ly/add-subtract-find-missing-value

Integer Addition & Subtraction Mystery Picture
Answer the problems correctly to reveal a hidden picture!
https://www.mathmammoth.com/practice/mystery-picture-integers#min=-50&max=50

Red and Black Triplematch Game for Adding Integers
This is a fun card game with 2-5 people to practice adding integers.
https://mathmamawrites.blogspot.com/2010/07/black-and-red-triplematch-card-game-for.html

Different Ways to Play "Integer War" Card Game
This page explains various ways to play a common card game that is used to review integers.
https://mathfilefoldergames.com/2013/05/13/integer-war/

Combine Like Terms with Negative Coefficients
Simplify algebraic expressions by combining like terms. Coefficients on some terms are negative.
https://bit.ly/Combine-Like-Terms

MULTIPLICATION AND DIVISION

Fruit Splat Multiplication of Integers
Click on the fruit that has the correct answer to the integer multiplication problem.
https://www.sheppardsoftware.com/math/integers/fruit-splat-multiplication/

Find the Missing Number Quiz
Practice integer multiplication with missing numbers in this interactive 10-question quiz.
https://www.thatquiz.org/tq-1/?-j114-la-p0

ALL OPERATIONS / GENERAL

Integers Bingo Game
Practice basic operations with integers (addition, subtraction, multiplication, or division) with this fun online bingo game!
https://www.mathmammoth.com/practice/bingo-integers

Rational Numbers on the Number Line
Practice placing positive and negative fractions and decimals on the number line in this interactive online activity.
https://bit.ly/Rational-Numbers-on-the-Number-Line

Integer Operations Quiz
Practice several operations with integers in this 10-question interactive quiz.
https://www.thatquiz.org/tq-1/?-jh8f-la-p0

Multiplying & Dividing Negative Numbers Word Problems (from Khan Academy)
Use this set of interactive word problems to reinforce your knowledge of integers.
https://bit.ly/multiply-divide-negative-numbers-problems

Order of Operations with Negative Numbers
Practice evaluating expressions using the order of operations in this interactive online activity.
https://bit.ly/Order-of-Operations-with-Negative-Numbers

Integers Quiz
Test your integer skills in this interactive online quiz.
https://www.thatquiz.org/tq/practicetest?10w790opx3kmd

Numerate Game with Integers
In this game, two players take turns forming equations using the available tiles.
https://www.transum.org/Maths/Game/Numerate/Default.asp?Level=2

Integers Jeopardy
A jeopardy game where the questions involve adding, subtracting, multiplying, and dividing integers.
https://www.math-play.com/Integers-Jeopardy/integers-jeopardy-fun-game_html5.html

Solve For Unknown Variable - Integer Review
Find the value of an unknown variable in a given addition or subtraction equation with integers.
https://maisonetmath.com/algebra/algebra-quizzes/294-solve-for-unknown-variable-integer-review

Fruit Shoot Game: Mixed Integer Operations
Practice all four operations with integers while shooting fruits. You can choose the difficulty level and the speed.
https://www.sheppardsoftware.com/math/integers/mixed-fruit-splat-game/

The History of Negative Numbers
While they seem natural to us now, in the past negative numbers have spurred controversy and have been called "fictitious" and other names.
https://nrich.maths.org/5961

Integers

The **counting numbers** are 1, 2, 3, 4, 5, and so on. They work for addition. But counting numbers do not allow us to perform all possible subtractions; for example, the answer to the problem 2 − 7 is not any of them. That is when we come to the *negatives* of the counting numbers: −1, −2, −3, −4, −5, and so on.

Together with zero, all these form the set of **integers**: {..., −4, −3, −2, −1, 0, 1, 2, 3, 4, ...}.
Note: Zero is neither positive nor negative.

Read −1 as "negative one" and −5 as "negative five." Some people read −5 as "minus five." That is very common, and it is not wrong, but be sure that you do not confuse it with subtraction.

Often, we need to put parentheses around negative numbers in order to avoid confusion with other symbols. Therefore, ⁻5, −5, and (−5) all mean "negative five."

Which is more, −30 or −5?

Which is *warmer*, −30°C or −5°C? Clearly −5°C is. Temperatures get colder and colder the more they move towards the negative numbers. We can write a comparison: −30°C < −5°C.

Similarly, we can write −$500 < −$200 to signify that to owe $500 is a worse situation than to owe $200.

Or, in elevation, −15 m > − 50 m means that 15 m below sea level is higher than 50 m below sea level.

1. Write comparisons using >, <, and integers. Don't forget to include the units!

   a. The temperature at 5 AM was 12°C below zero.
      Now, at 9 AM, it is 8°C below zero.

   b. I owe my mom $12, and my sister owes her $25.

   c. The bottom of the Challenger Deep trench is 11,033 m below sea level.
      Mt. Everest reaches to a height of 8,848 m.

   d. The total electric charge of five electrons is −5e.
      The total electric charge of 5 protons is +5e.
      (The symbol e means elementary charge, or a charge of a single proton.)

   e. Dean has $55, whereas Jack owes $15.

2. Which integer is ...

   a. 3 more than −7   b. 8 more than −3   c. 7 less than 2   d. 5 less than −11

3. Write the numbers in order from the least to the greatest.

a. −5 −56 51 −15	b. 3 0 −31 −13

41

The **absolute value** of a number is its distance to zero.

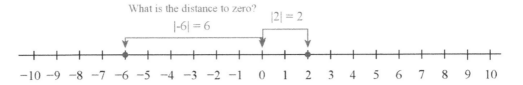

We denote the absolute value of a number by putting vertical bars on either side of it.
So $|-4|$ means "the absolute value of 4," which is 4. Similarly, $|87| = 87$. In an equation we treat the absolute value bars like parentheses and solve them first.

Example 1. Simplify $|-4| - |1|$. First simplify the absolute values. We get $4 - 1 = 3$.

Let's say someone's account balance is −$1,000. That person is in debt. The absolute value of the debt is written as $|-\$1000|$ and means that the *size* of the debt is $1,000.

If a diver is at a depth of −22 m, the absolute value $|-22 \text{ m}|$ tells us how far he is from the surface (22 m).

4. Simplify.

a. $|-11|$	b. $|+7|$	c. $|0|$	d. $|-46|$
e. $|-5| + |-2|$		f. $|-5| - |2|$	
g. $|-5| + |-2| + |8|$		h. $|5| + |-2| - |-1|$	

5. Answer. Use absolute values to calculate your answers.

   a. What is the distance between −153 and zero on a number line?

   b. What is the distance between *x* and zero on a number line?

   c. What is the distance between −11 and 21 on a number line?

6. Interpret the absolute value in each situation.

   a. A fishing net is at the depth of 15 feet. $|-15 \text{ ft}| = $ _____ ft

   Here, the absolute value shows _____

   b. The temperature is −5°C. $|-5°C| = $ _____ °C

   Here, the absolute value shows _____

   c. Eric's balance is −$7. $|-\$7| = \$$ _____

   Here, the absolute value shows _____

   d. A point is drawn in the coordinate grid at (0, −11) $|-11| = $ _____

   Here, the absolute value shows _____

The **opposite** of a number is the number that is on the opposite side of zero at the same distance from zero.

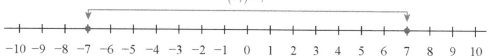

We denote the opposite of a number using the minus sign. For example, the opposite of 4 is written as −4. The opposite of −2 is written as −(−2), which is of course 2. So, −(−2) = 2.

The opposite of zero is zero itself. In symbols, −0 = 0.

"But wait," you might ask, "doesn't −4 mean 'negative four,' not 'the opposite of four'?"

It can mean either! Sometimes the context will help you tell which is which. Other times it isn't necessary to differentiate, because, after all, the opposite of four *is* negative four, or −4 = −4. ☺

In the expression −(4 + 5), the minus sign means the opposite of the <u>sum</u> 4 + 5, which equals negative nine.

Example 2. −|7| means the opposite of the absolute value of seven. It simplifies to −7.

Notice that there are *three* different meanings for the minus sign:

1. To indicate subtraction, as in 7 − 2.
2. To indicate negative numbers: "negative 7" is written −7.
3. To indicate the opposite of a number: −(n + 1) is the opposite of n + 1.

7. Write using symbols, and simplify if possible.

   a. the opposite of 6

   b. the opposite of −11

   c. the opposite of the absolute value of 12

   d. the absolute value of negative 12

   e. the opposite of the sum 6 + 8

   f. the opposite of the difference 9 − 7

   g. the absolute value of the opposite of 8

   h. the absolute value of the opposite of −2

8. Simplify.

   a. −|8| **b.** −(−9)   **c.** −|−7| **d.** −0   **e.** −(−(−100))

9. Write with symbols. Use a variable for the number.

   a. The sum of a number and five is greater than negative 6.

   b. The opposite of a certain number is less than negative 2.

   c. The absolute value of a certain number is greater than 15.

   d. The opposite of the quantity n plus 4 is less than minus 5.

10. Daniel owed $5. Then he borrowed $10 more. Next, he paid off $7 of his debts. Lastly he made yet another debt of $4. Write one integer to express Daniel's money situation now.

Remember **inequalities**? The number line below illustrates the inequality $x \geq -8$. Notice the arrow on the right, which shows that the ray continues to infinity.

The inequality $x > -4$ is plotted on the number line below. Notice the open circle, indicating that -4 is not part of the solution set.

11. Plot these inequalities on the number line. Don't forget the arrow on the open end.

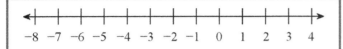
a. $x < -3$

b. $x > -1$

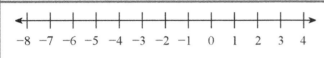
c. $x \geq -2$

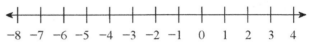
d. $x \leq 2$

12. **a.** Solve the inequality $x < 2$ in the set $\{-3, -2, -1, 0, 1, 2, 3\}$.

   b. Solve the inequality $x \geq -5$ in the set $\{-10, -8, -6, -2, -1, 0, 5\}$.

13. Write an inequality. Use negative integers where appropriate.

   a. The pit is at most 10 m deep.

   b. The pit is at least 12 m deep.

   c. Tim's debt is no more than $500.

   d. Nora owes at least $100.

   e. You have to have more than $20 to participate.

   f. For the skiing contest to take place, the temperature has to be warmer than 15 degrees below zero.

   g. The freezer temperature should be colder than 10 degrees below zero.

Puzzle Corner

Let a and b be two negative integers.
What is the distance between them on the number line?
Write an expression.

Addition and Subtraction on the Number Line

Addition can be modeled on the number line as a movement to the *right*.

Suppose you are at −4, and you jump 5 steps <u>to the right</u>. You end up at 1. We write the addition −4 + 5 = 1.

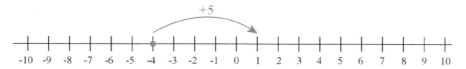

1. Draw a number line jump for each addition sentence and complete the addition sentence.

a. −8 + 3 = _____

b. −2 + 5 = _____

c. −4 + 4 = _____

d. −10 + 12 = _____

Subtraction can be shown on the number line as a movement to the *left*.

You are at −4, and you jump 5 steps <u>to the left</u>. You end up at −9. We write the subtraction −4 − 5 = −9.

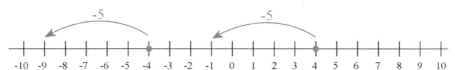

What subtraction does the other jump show? 4 − 5 = −1.

2. Draw a number line jump for each subtraction and complete the subtraction sentence.

a. 2 − 6 = _____

b. −2 − 5 = _____

c. 0 − 7 = _____

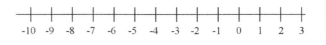

d. −5 − 4 = _____

When should you use parentheses around a negative number?

$-2 - 5 = -7$

$(-2) - 5 = (-7)$

$^-2 - 5 = {^-7}$

Use parentheses around a negative number if you need to make clear that the minus sign means "negative" and not subtraction. You can also use an elevated minus sign for clarity.

In the equation $-2 - 5 = -7$, there is no confusion, so the parentheses are not necessary.

3. Write an addition or a subtraction.   Addition/subtraction:

   a. You are at ⁻3. You jump 6 to the right. You end up at _____.

   b. You are at ⁻3. You jump 6 to the left. You end up at _____.

   c. You are at 2. You jump 7 to the left. You end up at _____.

   d. You are at ⁻10. You jump 9 to the right. You end up at _____.

   e. You are at ⁻7. You jump 4 to the right. You end up at _____.

4. Write an addition or subtraction to match the number line jumps.

   a.

   b.

   c.

   d.

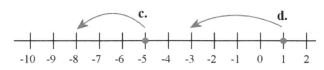

5. The temperature changes from what it was before. Find the new temperature.

before	2°C	0°C	1°C	−2°C	−12°C	−7°C
change	drops 3°C	drops 7°C	drops 5°C	rises 5°C	rises 6°C	rises 3°C
now						

6. Interpret each addition and subtraction in relation to a change in temperature.

   a. $-2 + 4 = 2$

   b. $-2 - 4 = -6$

   c. $-3 + 3 = 0$

Adding a negative integer

How would we show the sum 4 + (−2) on the number line?

Your starting point is 4. To add a negative number, we jump to the *left*, not to the right. (After all, 4 + (−2) cannot be the same as 4 + 2, so the only logical thing to do is to jump 2 in the *other* direction.)

So 4 + (−2) is really the same as 4 − 2!

Similarly, to show the sum −5 + (−3), we start at −5 and jump 3 steps to the left. It's the same as the subtraction −5 − 3.

In other words, when we add a negative number, the absolute value of the number gives us how many steps to jump: −2 + (−11) means, "Start at −2, and jump |−11| or 11 steps." Which direction? The minus sign means towards more negatives, or towards the left.

7. Draw a number line jump for each sum and complete the addition sentence.

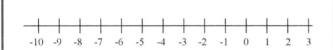

a. 2 + (−5) = _____

b. −2 + (−5) = _____

c. −1 + (−4) = _____

d. 3 + (−3) = _____

8. In the sum 7 + (−7), we are adding 7 to its opposite.

   a. Write another sum where you add a number and its opposite.

   b. Model your sum on the number line. Where do you end up?

   c. What is the result of adding any number *n* and its opposite −*n*?
   Write this as an equation.

9. Add or subtract. Think of the number line jumps.

a. 2 − 7 =	b. ⁻2 − 2 =	c. ⁻3 + 3 =	d. ⁻8 + 6 =
1 − 5 =	⁻6 − 3 =	⁻6 + 3 =	⁻12 + 4 =
5 − 9 =	⁻5 − 2 =	⁻15 + 5 =	⁻11 + 13 =

10. Fill in.

   a. For the sum −2 + 7, start at _____ and move _____ steps to the _____.

   b. For the sum −2 + (−7), start at _____ and move _____ steps to the _____.

11. Find the number that is missing from the equations. Think of jumps on the number line.

| a. 1 − _____ = ¯2 | c. ¯7 + _____ = ¯4 | e. 1 − _____ = ¯6 | g. ¯5 + _____ = 0 |
| b. 3 − _____ = ¯5 | d. ¯9 + _____ = ¯4 | f. 0 − _____ = ¯9 | h. ¯7 + _____ = 7 |

12. The expression 1 − 3 − 5 − 7 can also be thought of as a person making jumps on the number line. Where does the person end up?

13. James had $10. He bought coffee for $2 and a sandwich for $12. He paid what he could and the clerk put the rest of the bill on his charge account with the store. Write an equation to show the transaction.

14. Interpret the sum −2 + (−5) in the context of money. Then solve it.

15. Interpret the sum 2 + (−5) in the context of money. Then solve it.

What about subtracting a negative number?

Here is a way to think about 3 − (−2) on the number line. Imagine you are standing at 3. Because of the subtraction sign, you turn to the *left* and get ready to take two steps. However, because of the *additional* minus sign in front of the 2, you have to take those steps BACKWARDS—to the *right*! So, because you ended up taking those 2 steps to the right, in effect you have just performed 3 + 2!

Quite surprising, but true: that double minus sign in 3 − (−2) turns into a plus!

16. Draw a number line jump for each subtraction, using the method just explained.

a. 1 − (−2) = _____

b. 0 − (−2) = _____

c. −9 − (−4) = _____

d. −3 − (−5) = _____

Addition of Integers

Addition of integers can be modeled using **counters.** We will use green counters with a "+" sign for positives and red counters with a "−" sign for negatives.

This picture shows the sum $(-2) + (-3)$. We *add* negatives and negatives. In total, there are five negatives, so the sum is -5.	 $1 + (-1) = 0$ One positive counter and one negative counter *cancel* each other. In other words, their sum is zero!	$(-4) + 3 = -1$ The negatives outweigh the positives. Pair up three negatives with three positives. Those cancel each other out. There is still one negative left.

1. Refer to the pictures and add. Remember that each positive-negative pair is canceled.

a. $3 + (-1) =$ _____	**b.** $(-4) + (-4) =$ _____	**c.** $2 + (-4) =$ _____

2. Write addition equations to match the pictures.

a.	**b.**	**c.**

d.	**e.**	**f.**

Reminder. Use parentheses if plus and minus signs, or two minus signs, are next to each other $(+-, -+, --)$. So, instead of "$2 + - 4$," write "$2 + (-4)$." And instead of "$2 - -3$," write "$2 - (-3)$."

3. Think of the counters. Add.

a. $2 + (-8) =$	**b.** $(-2) + (-8) =$	**c.** $3 + (-10) =$	**d.** $15 + (-20) =$
$(-2) + 8 =$	$2 + 8 =$	$10 + (-3) =$	$20 + (-50) =$
e. $-5 + (-4) =$	**f.** $18 + -1 =$	**g.** $-9 + 2 =$	**h.** $-20 + (-5) =$
$-6 + 6 =$	$-18 + (-1) =$	$-9 + (-2) =$	$20 + (-5) =$

4. Compare how $-8 + 6$ is modeled on the number line and with counters.

   a. On the number line, $-8 + 6$ is like starting at _____, and moving _____ steps to the _____, ending at _____.

   b. With counters, $-8 + 6$ is like _____ negatives and _____ positives added together. We can form _____ negative-positive pairs that cancel each other out, and what is left is _____ negatives.

5. Add. You can think of counters or number line jumps.

a. $2 + (-11) =$	b. $-11 + (-11) =$	c. $-2 + (-9) =$	d. $21 + (-7) =$
$-7 + 9 =$	$3 + (-8) =$	$16 + (-5) =$	$-30 + 20 =$

6. Write an addition equation for each situation.

   a. Your checking account is overdrawn by $30. (This means your account is negative.) Then you deposit $50. What is the balance in your account now?

   b. A chlorine atom has 17 protons in its nucleus and 17 electrons surrounding the nucleus. (It also has some neutrons which have no electric charge.) The electric charge of the protons is $17e$, and the charge of the electrons is $-17e$. What is the total electric charge of this atom?

   c. A chlorine ion has 17 protons and 18 electrons. The electric charge of the protons is $17e$, and the charge of the electrons is $-18e$. What is the total electric charge of this ion?

7. Draw a picture using counters to model the addition of the number -6 and its opposite. What is the sum?

8. a. Find the value of the expressions $p + 3$ and $p + (-6)$ for the different values of p.

p	$p + 3$	$p + (-6)$	p	$p + 3$	$p + (-6)$	p	$p + 3$	$p + (-6)$
-6			-2			2		
-5			-1			3		
-4			0			4		
-3			1			5		

   b. Consider the patterns in the values of these expressions. Is there any value for p for which $p + 3$ would be equal to $p + (-6)$?

   c. For which values in the table is $p + 3$ more than $p + (-6)$?

   d. For which value of p does the expression $p + (-4)$ have the value 0?

How do we solve 84 + (−54)?	**How do we solve −55 + 23?**
There are 84 positives and 54 negatives, so obviously the positives will "win." We can make 54 positive-negative pairs that cancel each other out.	Since there are 55 negatives and 23 positives, the negatives will "win." We can make 23 positive-negative pairs that cancel each other out.
The <u>difference</u> of 84 and 54 (which is 30) tells us how many positives are "left over" after canceling. That is why 84 + (−54) = 30.	The <u>difference</u> of 55 and 23 (which is 32) tells us how many negatives are "left over" after canceling. Therefore, −55 + 23 = −32.
You could also use the shortcut where the "+ −" in the middle is changed to a − : 84 + (−54) = 84 − 54 = 30	There is no shortcut as such for this problem. However, notice that even though it is an *addition* problem, we end up solving it by *subtracting* (because we have to find the difference)!

To add a positive number and a negative number, calculate the difference between their absolute values. That difference tells you by how much the majority side "wins."

The final sum is negative if there were more negatives to start with and positive if there were more positives.

9. Fill in.

   a. To solve (−92) + 31, first check: Are there more positives or negatives? _____

   Find the difference between the absolute values, 92 and 31: _____. So the final answer is _____.

   b. To solve 92 + (−31), first check: Are there more positives or negatives? _____

   Find the difference between the absolute values, 92 and 31: _____. So the final answer is _____.

   c. To solve (−92) + (−31), notice that both addends are _____. Therefore, you _____

   the absolute values, 92 and 31: _____. So the final answer is _____.

10. Solve.

a. 22 + (−33) =	**d.** 50 + (−59) =	**g.** 72 + (−60) =
b. (−60) + 45 =	**e.** −50 + (−59) =	**h.** −51 + 21 =
c. (−100) + 70 =	**f.** −50 + 59 =	**i.** (−15) + (−37) =

11. A plutonium ion has 94 protons and 90 electrons. The electric charge of the protons is $94e$, and the charge of the electrons is $-90e$. What is the total electric charge of this ion?

12. Consider the four expressions 72 + 16, (−72) + (−16), (−72) + 16, and 72 + (−16). Write these expressions in order from the one with **least** value to the one with **greatest** value.

Subtraction of Integers

Subtraction using counters

Using counters, subtraction is modeled as "taking away." So 6 − 4 means we start with 6 positive counters and take away 4 of them.

⁻5 − (⁻3) means we start with 5 negative counters, and then take away 3 negative counters. That leaves 2 negative counters, or ⁻2.

⁻5 − (⁻3) = ⁻2

Yet we cannot easily model 5 − (⁻4) like that because we cannot take away 4 negative counters when there are none! But we could do it this way:

Start out with 5 positives. Then add four positive-negative *pairs*, which is the same as adding zero! That won't affect the net value of the counters at all.

Now we *can* take away the four negatives. We're left with nine positives: 5 − (⁻4) = 9.

Notice again that the double negative works out to be the same as a single plus sign: 5 − (⁻4) is the same as 5 + 4.

 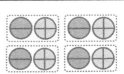

(1) Start with five positives.
(2) Add four positive-negative pairs, which amount to zero.
(3) Cross out the four negatives.

Nine positives are left: 5 − (⁻4) = 9

To model ⁻6 − 2 with counters, we start out with six negatives. But we cannot take away two positives!

So, we add zero, in the form of two positive-negative *pairs*. Adding zero does not affect the net value of the counters.

Now we can take away two positives. We're left with eight negative counters: ⁻6 − 2 = ⁻8.

(1) Start with six negatives.
(2) Add two positive-negative pairs (a zero).
(3) Now cross out the two positives.
Eight negatives are left: ⁻6 − 2 = ⁻8.

1. Model the subtractions with counters. You may need to add positive-negative pairs before subtracting.

a. ⁻4 − (⁻2) = _____	**b.** 4 − (⁻2) = _____	**c.** ⁻3 − 4 = _____	**d.** 3 − (⁻2) = _____

2. Now draw the counters yourself.

a. 2 − (⁻3) = _____	**b.** ⁻7 − (⁻5) = _____	**c.** ⁻3 − 6 = _____

Addition changed to subtraction and vice versa

We can think of $-5 + (-3)$ as five negatives and three more negatives, totaling eight negatives, or -8. Also, $-5 - 3$ is like starting at -5 and jumping three steps to the left on the number line, ending at -8.

Since both expressions have the same answer, they are equal:

$$-5 + (-3) = -5 - 3$$

We see that adding -3 is the same as subtracting 3. And it's as if the "+ −" in the middle were changed into a single "−" sign. (That's a shortcut!)

If we turn this same equation around, we get

$$-5 - 3 = -5 + (-3)$$

You can think of it like this: **Subtraction can be changed into the addition of the opposite.**

3. Is that really true? Try it yourself. In (a) and (b) check that you get the same answer. In (c) and (d) write the matching addition or subtraction.

a. $2 - 7 = $ _____	b. $-1 - 8 = $ _____	c. $3 - 6 = $ _____	d. ___ − ___ = ___
$2 + (-7) = $ _____	$-1 + (-8) = $ _____	___ + ___ = ___	$-7 + (-5) = $ _____

But *wait*! What about an addition like $2 + 4$? Can that also be changed into subtraction? Or can $9 - 1$ be changed into an addition? Or $-3 - (-2)$?

Yes, they can.

Remember when you saw a double negative turn into an addition? For example, to solve $-5 - (-3)$ on a number line, you start at -5, head left because of the subtraction, but then because of the -3, you go backwards and end up moving *right*! You land at -2, and the expression $-5 - (-3)$ ends up being equal to $-5 + 3$.

$$-5 - (-3) = -5 + 3$$

or

$$-5 + 3 = -5 - (-3)$$

Any subtraction can be changed into an addition, *and* vice versa.

Instead of subtracting a number, you can add its opposite.
Instead of adding a number, you can subtract its opposite.

Example. You can change the addition $2 + 8$ into a subtraction: instead of adding 8, subtract its opposite, which is -8. So, $2 + 8$ is equal to $2 - (-8)$. This change did not make it easier to solve, but it is true anyway.

4. Write the matching subtraction or addition and solve it.

a. $2 - (-7)$	b. $7 + (-3)$	c. $-1 - (-4)$	d. $2 + (-5)$
↓	↓	↓	↓
$2 + 7 = $ _____	___ ___ = ___	___ ___ = ___	___ ___ = ___

> Mathematicians actually define subtraction of two numbers, $a - b$, as the <u>sum</u> of a and the opposite of b.
>
> In symbols: $a - b = a + (-b)$
>
> From this definition it also follows that $a - (-b)$ simplifies to $a + b$.
>
> (Why? In $a - (-b)$ we subtract $-b$, and the opposite of $-b$ is b. Instead of subtracting $(-b)$, you *add* its opposite, or b.)

Examples. Here I will explain how I (Maria) think about integer additions and subtractions.

$-2 + (-9)$	$-2 - (-9)$	$-2 - 9$
I think of it as negatives and more negatives.	I change this into the addition $-2 + 9$.	I view this as having some negatives, and then moving on the number line towards more negatives.

$-2 + 9$

I think of this as a a number line jump, starting at -2 and jumping 9 steps, ending at 7.

Or I might turn it around and think of it as $9 - 2$.
Why? Because addition is commutative: $-2 + 9 = 9 + (-2)$, and then that is equal to $9 - 2$.

5. Change each addition into a subtraction or vice versa. Then solve whichever is easier. Sometimes changing the problem will not make solving it easier, but the aim of this exercise is to practice making the change!

a. $-5 + (-3)$ ↓ ___ ___ = ___	b. $10 - (-3)$ ↓ ___ ___ = ___	c. $-1 + (-4)$ ↓ ___ ___ = ___	d. $2 + 6$ ↓ ___ ___ = ___
e. $-11 - (-6)$ ↓ ___ ___ = ___	f. $0 + (-7)$ ↓ ___ ___ = ___	g. $9 + 8$ ↓ ___ ___ = ___	h. $-2 - 9$ ↓ ___ ___ = ___

6. Think up a real-world context for each calculation.

   a. -20 ft $- 30$ ft

   b. $\$15 - \$25$

7. **a.** Consider the expression $2 - x$. Will this expression have a greater value when x has the value -2 or when x is -5?

   b. What about when x is 3 or when x is 7?

Shortcuts for simplification:
− − can be changed to a single plus sign: $n - (-m) = n + m$
+ − can be changed to a single minus sign: $n + (-m) = n - m$
− + can be changed to a single minus sign: $n - (+m) = n - m$

8. Apply the shortcuts from the box above to simplify these expressions by removing the parentheses.

a. $2 + (-s) =$	b. $5 - (-x) =$	c. $x + (-y) =$
d. $2t - (-3s) =$	e. $8 - (+7x) =$	f. $7m + (-8p) =$
g. $2x - (-y) =$	h. $6w + (-1.3) =$	i. $5x^2 - (-9) =$

9. Simplify these, too!

a. $2x - (-x) =$	b. $5y + (-y) =$	c. $7w + (-2w) =$

10. Solve and continue the patterns to answer the questions.

a. $^-7 - 3 =$	b. $-5 + 3 =$	c. $2 - (-3) =$
$^-7 - 2 =$	$-5 + 2 =$	$2 - (-2) =$
$^-7 - 1 =$	$-5 + 1 =$	$2 - (-1) =$
$^-7 - 0 =$	$-5 + 0 =$	
$^-7 - {}^-1 =$		
For what value of x will $-7 - x$ have the value of 2?	For what value of x will $-5 + x$ have the value of -12?	For what value of x will $2 - x$ have the value of -7?

11. **a.** Find the value of the expression $-4 - x$ for at least six different values of x. Make a table to organize your work. You should see a pattern if you choose the values of x that are in some order.

   b. For which value of x will the expression $-4 - x$ have a value of 0?

Adding or Subtracting Several Integers

Let's solve $(-3) + 4 + (-8) + 5$. It makes sense to add all the negatives and all the positives together first.

We get $-3 + (-8) = -11$ for the negatives, and $4 + 5 = 9$ for the positives. Lastly, we combine those, and get $-11 + 9 = -2$.

You *could* have added the numbers in order from left to right: First $(-3) + 4 = 1$, then $1 + (-8) = -7$, and lastly $-7 + 5 = -2$, but that method is a lot slower when you need to add a lot of integers.

Example. $5 + (-2) + 7 + (-8) + 1 = ?$

First add all the negatives: $(-2) + (-8) = -10$. Then add all the positives: $5 + 7 + 1 = 13$. Lastly add -10 and 13: $-10 + 13 = 3$.

When you have three or more integers to add:

1) First add all the negatives and all the positives separately.

2) Finally add the two sums you obtained in step (1).

1. Add.

a. $(-4) + 5 + 6 + (-1) = $ _____	**b.** $-2 + (-5) + 10 + (-7) = $ _____
c. $8 + (-2) + (-6) + (-3) = $ _____	**d.** $6 + (-3) + 8 + (-3) = $ _____

2. Add and solve the riddle. *What building has the most stories?*

A. $5 + {}^-4 + {}^-1 = $ _____	**Y.** $12 + {}^-2 + {}^-2 + {}^-2 = $ _____	**E.** $^-7 + {}^-8 + {}^-2 = $ _____
I. $^-7 + {}^-7 + 5 = $ _____	**T** $^-10 + 6 + {}^-5 + 5 = $ _____	**H.** $10 + {}^-2 + 15 = $ _____
L. $^-5 + 2 + 8 = $ _____	**R.** $5 + 1 + {}^-11 = $ _____	**B.** $^-5 + 8 + {}^-11 = $ _____

3. Solve $(-15) + (-6) + 9 + 12 + (-4) + 2$.

4. Complete the equations. There are many possible solutions. Find at least two different ones!

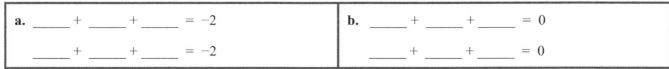

> **How to add and subtract several integers**
>
> 1) First change all subtractions into additions.
>
> 2) Add the integers using the method explained in the previous page.
>
> **Example.** $5 - (-2) + 7 - 8 + 1 = ?$
>
> There are two subtractions. Changing those into additions, we get $5 + 2 + 7 + (-8) + 1 = ?$
>
> Now, we add all the positives: $5 + 2 + 7 + 1 = 15$. There is only one negative integer, -8. Lastly, we add $15 + (-8) = 15 - 8 = 7$.

5. Add and subtract. First, change each subtraction into an addition. The answers are on the right.

a. $-12 - (-2) - (-5) =$	**b.** $6 - (-1) + (-5) =$
c. $-9 - (-8) - (-2) + 4 =$	**d.** $6 + (-1) - (-5) - 3 =$
e. $-8 + (-2) - (-6) - 8 =$	**f.** $6 + (-1) - (-3) - 7 =$

5
-12
2
7
-5
1

6. Match the equations with the situations and complete the missing parts.

   a. John had a debt of $8. He earned $6. Then he earned $12 more.
   Now he has _____.

   b. A diver was at the depth of 8 ft. Then he rose 6 ft. Then he sank 12 ft.
   Now he is at _____ ft.

   c. The temperature was $-8°C$ and fell 6°. Then it rose 12°.
   Now the temperature is _____ °C.

   $-8 - 6 + 12 =$ _____
   $-8 + 6 - 12 =$ _____
   $-8 + 6 + 12 =$ _____

7. Write an expression to match the situation, and simplify it.

   Hannah owed $10 to her mom. Then, she borrowed $10 more from her mom to buy a T-shirt. Lastly, she earned $50, and paid her mom back. What is Hannah's balance now?

8. Write a story about money to match the expression $-2 - (-10) + (-7) - 4$.

Distance and More Practice

1. **a.** How far apart are 55 and 91?

   b. How about −55 and −91?

   c. What is the distance between 1,091 and 342 on a number line?

   d. How far apart are −91 and 32 on a number line?

2. **a.** Explain how you can find the distance between any two *positive* numbers a and b.
   Also, write your method using symbols and variables, if you can.

   b. Explain how you can find the distance between any two *negative* numbers.
   Also, write your method using symbols and variables, if you can.

You probably used the *difference* of two numbers to find the distance between them:

$$\text{The difference of } a \text{ and } b \text{ is } a - b.$$

However, for that to be the <u>distance</u> between a and b, you need to subtract the smaller number from the bigger number, or you might get a negative answer—and distance is never negative!

For example, to find the distance between 475 and 1,091, you need to subtract $1{,}091 - 475$, not $475 - 1{,}091$.

Just to write $a - b$ for the distance between a and b is not enough. It is, if $b \le a$. But if $b > a$, it will give us a negative number, the opposite of the correct result.

We do have a way around this, and that is to take the **absolute value of the difference**. Remember, an absolute value is never negative. If the difference of two numbers is positive, its absolute value is the number itself, but if the difference is negative, you drop the negative sign. That is exactly what we want!

$$\text{The distance between } a \text{ and } b \text{ is } |a - b|.$$

This formula for distance works with negative numbers as well, as you will soon see!

---

3. Which expressions can be used to find the distance between 16 and 9?

a. $9 - 16$	b. $16 - 9$	c. $|9 - 16|$	d. $|16 - 9|$	e. $|-16 - 9|$

4. Explain how you can find the distance between a positive number and a negative number, such as 87 and −92. Also, write your method using symbols and variables, if you can.

> The distance between a and b is $|a-b|$.

Example 1. What is the distance between -2 and -11? You can probably easily see it is 9 units (by realizing that it is the same as the distance between 2 and 11).

Will the formula above work for finding the distance? Notice carefully how we put the numbers into the formula. We find that the distance is $|-2-(-11)|$.

Now, to simplify it, first change the double negative into addition: $|-2-(-11)| = |-2+11|$. Then calculate what is inside the absolute value bars: $-2 + 11 = 9$. Lastly, we have $|9|$, which equals 9.

So the formula did work, but it was quicker to calculate the distance without it.

Example 2. What is the distance between 5 and -19? You can probably see it is $5 + 19 = 24$ units. But will the formula above work for finding the distance?

Notice carefully how we put the numbers into the formula. We find that the distance is $|5-(-19)|$.

Now, to simplify it, first change the double negative into addition: $|5-(-19)| = |5+19|$. Then calculate $5 + 19 = 24$. Lastly, we have $|24|$, which equals 24.

So, once again, the formula did work, but it was quicker to calculate the distance without it.

Now you might ask, "What is the good of that formula if it is quicker not to use it?"

Using the simpler method works fine with numbers. But what if you need to know the distance between a and b? Which is bigger? Is the distance $a - b$, or is it $b - a$? You don't know. Absolute value notation gives you a way to say, "Take the difference, and if it happens to come out negative, then make it positive." That is a really handy tool.

5. Evaluate the expression $|a-b|$ for the given values of a and b. Check that the answer you get is the same as if you had used a number line to figure out the distance between the two numbers.

a. a is 7 and b is 92 $\|7-92\| =$	**b.** a is -3 and b is 5
c. a is -8 and b is -5	**d.** a is 7 and b is -5
e. a is 0 and b is 14	**f.** a is -2 and b is -9

6. What happens if you calculate the distance between two numbers m and n as $|n-m|$ instead of $|m-n|$? In other words, what happens if you reverse the order of the numbers in the subtraction?

Investigate this by checking several pairs of numbers. For example, what happens if you calculate the distance between 18 and 11 as $|11-18|$ instead of $|18-11|$?

Try some pairs of negative numbers, as well. Then try some pairs where one number is positive and the other is negative. Does the formula still work?

You can *also* calculate the distance between *a* and *b* as $|b - a|$. The order in which you subtract the two numbers doesn't matter because, in the end, taking the absolute value makes the answer positive.

7. Which expressions can be used to find the distance between −6 and 9?

| a. $9 - 6$ | b. $|6 - 9|$ | c. $-6 - 9$ | d. $|9 - 6|$ | e. $|9 - (-6)|$ | f. $|-6 - 9|$ |

8. Which expressions can be used to find the distance between −5 and −11?

| a. $-11 - 5$ | b. $|-11 + (-5)|$ | c. $|-5 - (-11)|$ | d. $5 - 11$ | e. $|-11 - 5|$ | f. $|-11 - (-5)|$ |

9. Which expressions can be used to find the distance between *x* and 8?

| a. $x - 8$ | b. $|x + 8|$ | c. $x - (-8)$ | d. $|8 - x|$ | e. $|x - 8|$ | f. $|x - (-8)|$ |

10. Which expressions can be used to find the distance between *x* and −12?

| a. $x - 12$ | b. $|x - 12|$ | c. $x - (-12)$ | d. $|-12 - x|$ | e. $|x - (-12)|$ | f. $12 - x$ |

11. How much more money does Alex have Amy if Amy's balance is −$28 and Alex's is $45?

12. Augustus Caesar reigned as the emperor of Rome from 27 BC to 14 AD, and his successor Tiberius reigned from 14 AD to 37 AD. How much longer did Augustus reign than Tiberius? Note: There is no year zero in the calendar. The years jump from 1 BC directly to 1 AD. (So how will you account for that jump in your solution?)

13. Give an example where one person's account balance is $30 better than another's, yet both have a negative balance.

14. The table lists the average high and low temperatures, measured over a 30-year period, for each month in Río Grande, Argentina, which is located very near the southern tip of South America. For each month, calculate the temperature difference between the average low and the average high temperatures.

Río Grande	JAN	FEB	MAR	APR	MAY	JUN	JUL	AUG	SEP	OCT	NOV	DEC
Average High Temperature (°C)	16.2	15.9	13.6	10.2	6.3	3.0	2.9	5.1	8.5	11.5	13.8	15.4
Average Low Temperature (°C)	5.6	5.3	3.3	1.1	−0.9	−3.0	−2.6	−1.9	−0.4	1.7	3.4	4.8
Temperature difference												

What is the warmest month? _____ The coldest month? _____

How much does the temperature difference change between the warmest month and the coldest month? _____

15. Find the numbers that are missing from the equations.

a. −3 + _____ = −7	b. −3 + _____ = 3	c. 3 + _____ = (−7)
d. _____ + (−15) = −22	e. 2 + _____ = −5	f. _____ + (−5) = 0

16. Solve (−9) + 18 + (−2) + (−5) + 9.

17. Allison's mom designed a reward system for Allison where she would get positive points for chores and school work well done and negative points for chores and school work poorly done.

    Here is her list of points for one week. Calculate Allison's "total" for the week.

	positives	negatives
Mon	12	6
Tue	10	8
Wed	7	10
Thu	11	5
Fri	9	2
Sat	12	5

18. Explain a real-life situation for the addition −20 + 20 = 0.

19. Write an equation to match each situation.

    a. Mary had a debt of $30. She earned $10. Then she earned $15 more. Now she has _____.

    b. A diver was at a depth of 10 ft. Then he sank 5 ft. Then he sank 15 ft more. Now he is at the depth of _____ ft.

    c. The temperature was 2°C and fell 7°. Then it rose 4°. Now the temperature is _____ °C.

20. Let $x = -3$ and $y = 7$. Evaluate the expressions.

a. $x + y$	b. $x - y$	c. $y - x$	d. $|y - x|$

a. Consider the expressions $x − 6$ and $x + 6$.
Is $x − 6$ always less than $x + 6$, no matter what value you give to x? Study this by trying different values for x, including negative numbers.

b. Do the same with $5 − x$ and $5 + x$. Is it always true that $5 − x < 5 + x$?

Multiplying Integers

Multiply a positive times a negative

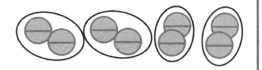

The image illustrates 4 · (−2) as four groups of two negatives. We can solve it using repeated addition:

4 · (−2) = (−2) + (−2) + (−2) + (−2) = −8.

As a shortcut, just multiply 4 · 2, and write the answer as a negative.

Example. 7 · (−8) = ?

This is illustrated by 7 groups of 8 negatives, which means the answer will be negative.
We multiply 7 · 8 = 56 to find how many negatives there are. The final answer is 7 · (−8) = −56.

1. Multiply.

a. 5 · (−4) = _____	**b.** 8 · (−1) = _____	**c.** 9 · (−9) = _____
12 · (−2) = _____	7 · (−6) = _____	10 · (−7) = _____

2. Write each addition as a multiplication, and solve.

a. ⁻4 + ⁻4 + ⁻4 + ⁻4 = ____ · ____ = ____	**b.** ⁻31 + ⁻31 = ____ · ____ = ____	**c.** ⁻200 + ⁻200 + ⁻200 = ____ · ____ = ____

Multiply a negative times a positive

Because multiplication is commutative, to solve a negative times a positive, like (−8) · 4 or −5 · 6, we can change the order of the factors to positive times negative.

(−8) · 4 is the same as 4 · (−8) = −32.

−5 · 6 is the same as 6 · (−5) = −30.

So not only does a positive times a negative make a negative, but a negative times a positive also makes a negative. However, notice that −5 · 0 = 0. Zero isn't written as −0, but always as 0.

3. Multiply.

a. −5 · 7 = _____	**b.** (−9) · 1 = _____	**c.** (−9) · 0 = _____
11 · (−3) = _____	−8 · 8 = _____	8 · (−5) = _____

4. What about a negative times negative? What do you think −2 · (−3) should be?

The answer that mathematicians have come up with is on the next page.

Multiply a negative times a negative

What is $(-8) \cdot (-4)$ or $-5 \cdot (-6)$?

This has baffled even professional mathematicians in the past, too, so don't worry if the answer at first seems confusing!

A negative times a negative makes a *positive*!

(Why? We will explore that in Exercise 5 below.)

So $(-8) \cdot (-4) = 32$, and $-5 \cdot (-6) = 30$.

5. Complete the patterns.

a.	b.	c.
$-3 \cdot 3 =$ _____	$-5 \cdot 3 =$ _____	$(-8) \cdot 3 =$ _____
$-3 \cdot 2 =$ _____	$-5 \cdot 2 =$ _____	$(-8) \cdot 2 =$ _____
$-3 \cdot 1 =$ _____	$-5 \cdot 1 =$ _____	$(-8) \cdot 1 =$ _____
$-3 \cdot 0 =$ _____	$-5 \cdot 0 =$ _____	$(-8) \cdot 0 =$ _____
$-3 \cdot (-1) =$ _____	$-5 \cdot (-1) =$ _____	$(-8) \cdot (-1) =$ _____
$-3 \cdot (-2) =$ _____	$-5 \cdot (-2) =$ _____	$(-8) \cdot (-2) =$ _____
$-3 \cdot (-3) =$ _____	$-5 \cdot (-3) =$ _____	$(-8) \cdot (-3) =$ _____
$-3 \cdot (-4) =$ _____	$-5 \cdot (-4) =$ _____	$(-8) \cdot (-4) =$ _____
In the pattern above, the products (answers) increase by 3 in each step!	In the pattern above, the products (answers) increase by ____ in each step!	In the pattern above, the products (answers) increase by ____ in each step!

The patterns in the products show that, to be consistent, a negative times a negative must be a positive.

6. Multiply.

a. $-5 \cdot 4 =$ _____	b. $(-9) \cdot (-2) =$ _____	c. $(-3) \cdot 30 =$ _____
$-5 \cdot (-4) =$ _____	$2 \cdot (-11) =$ _____	$-7 \cdot (-80) =$ _____

7. **a.** Margie bought four different colors of paint for $6 each. However, she did not have any money, so she bought them on credit. Write an integer *product* to describe her balance.

b. The sulfate ion consists of a central sulfur atom surrounded by four oxygen atoms. Its chemical formula is SO_4, and its electrical charge is $-2e$. What is the total charge on six sulfate ions?

c. A positive hydrogen ion has an electric charge of $+1e$. How many sulfate ions would you need to neutralize 24 hydrogen ions (to bring the total electrical charge to zero)?

A justification for the rule "Negative times negative makes positive"

This justification uses the distributive property, which states that $a(b + c) = ab + ac$.

Let's see what happens if $a = -1$, $b = 3$, and $c = -3$. We get

$$(-1) \cdot [3 + (-3)] = (-1) \cdot 3 + (-1) \cdot (-3)$$

Now, since $3 + (-3)$ on the left side is zero, the product on the left side is zero!

$$0 = (-1) \cdot 3 + (-1) \cdot (-3)$$

This means that the right side *must* be zero as well! On the right side, $(-1) \cdot 3$ equals -3. So the only way to make the right side also equal zero is for $(-1) \cdot (-3)$ to be $+3$.

Therefore, $(-1) \cdot (-3)$ equals *positive 3*.

We want mathematics be consistent without exceptions. In order for the patterns that we saw on the last page, the distributive property above, and lots of other procedures to give us consistent answers, we need to agree that "negative times negative is positive." Then everything works out fine.

8. Try this same argument yourself!

   (1) Substitute $a = -1$, $b = 1$, and $c = -1$ into the formula for the distributive property $a(b + c) = ab + ac$.

   ____ (____ + ____) = ____ · ____ + ____ · ____

   (2) The whole left side is zero because ____ + ____ = 0.

   (3) So the right side must equal zero as well.

   (4) On the right side, $-1 \cdot 1$ equals _____. Therefore, $-1 \cdot (-1)$ must equal _____ so that the sum on the right side will equal zero.

   (5) Therefore, $-1 \cdot (-1)$ must equal _____ .

Now let's consider $(-2) \cdot (-1)$. Since -2 is equal to $2 \cdot (-1)$, we can write $(-2) \cdot (-1) = 2 \cdot (-1) \cdot (-1)$ Since multiplication is associative, we can compute $(-1) \cdot (-1)$ first. But in exercise 8 we just proved that $(-1) \cdot (-1) = 1$.	So we see that $2 \cdot (-1) \cdot (-1) = 2 \cdot 1 = 2$ You can repeat this same argument replacing -2 with any negative number you wish. Thus, the product of any negative number and -1 is positive. In general, multiplying any number by negative one changes it into its opposite. In symbols, $m \cdot (-1) = -m$.
Lastly, we can use an argument similar to the one above to prove that the product of any negative number times any negative number will be positive.	For example: $(-5) \cdot (-3)$ $= 5 \cdot (-1) \cdot (-1) \cdot 3$ $= 5 \cdot 1 \cdot 3 = 5 \cdot 3$

At these websites you can read about the difficulties mathematicians have had in accepting negative numbers.

The History of Negative Numbers: https://nrich.maths.org/5961

9. Find the missing factors.

a. 4 · _____ = −32	b. −9 · _____ = 108	c. 9 · _____ = −900
d. −4 · _____ = 32	e. −9 · _____ = −108	f. −9 · _____ = 900

10. Multiply, and solve the riddle.

What is black when it is clean, and white when it is dirty?

A. 5 · (−4) = _____	D. (−4) · 9 = _____	C. 6 · (−5) = _____
K. −7 · (−7) = _____	O. 7 · (−7) = _____	R. −3 · (−8) = _____
A. 2 · (−12) = _____	L. (−4) · (−5) = _____	B. (−3) · (−12) = _____
	A. −3 · (−10) = _____	B. (−2) · (−5) = _____

−20   10   20   −24   −30   49   36   −49   30   24   −36

11. Think up a real-life situation for the product 3 · (−10).

12. The points (−2, 1), (0, 0), and (1, 2) are vertices of a triangle.

   a. Draw the triangle.

   b. Multiply each coordinate of each point by 3, to get three new points.
      Write the coordinates of the new points:

   c. Draw a new triangle using the three new points.

   What you just did was *enlarge* the original triangle. The original and the new triangle are *similar triangles*—they have the same shape.

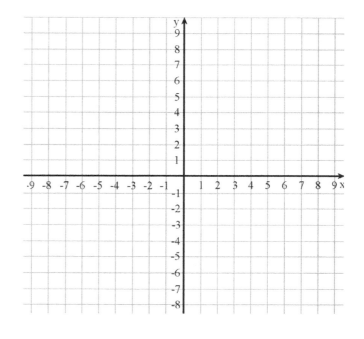

13. a. Draw a quadrilateral with vertices (−1, −1), (2, −1), (3, −3), and (−4, −3).

    b. Enlarge the quadrilateral by multiplying all the coordinates by 2. Draw the resulting quadrilateral.

Multiplying more than two integers

To solve $(-7) \cdot 2 \cdot (-2)$, we could multiply in order from left to right. First, $(-7) \cdot 2 = -14$, and then $-14 \cdot (-2) = 28$.

However, there is a shortcut. You can multiply the absolute values (the "plain" unsigned numbers) first, and then figure out if your answer will be positive or negative by counting the negative factors:

- If there is an *odd* number of negative factors, the answer is *negative*.
- If there is an *even* number of negative factors, the answer is *positive*.

Why is it so? Every time you have an even number of negative factors, you can pair them two by two, and each pair, when multiplied, produces a positive number. You end up multiplying positive numbers.

So what happens when there is an odd number of negative factors? Think for yourself.

Example. Multiply $4 \cdot (-2) \cdot (-1) \cdot 10 \cdot (-3)$.

First multiply $4 \cdot 2 \cdot 1 \cdot 10 \cdot 3 = 8 \cdot 10 \cdot 3 = 240$. Now count the negative factors. There are three: -2, -1, and -3. Three is an odd number, so the answer is negative: -240.

14. Edith is finding the value of the expression $(-5) \cdot (-2) \cdot (-1) \cdot 4 \cdot (-10)$. Complete her reasoning.

    "First I multiply the unsigned numbers: $5 \cdot 2 \cdot 1 \cdot 4 \cdot 10 = $ _____. Then, I count how many negative factors there are. There are _____ negative factors. That is an (odd/even) number, so the final answer is (positive/negative). Therefore, $(-5) \cdot (-2) \cdot (-1) \cdot 4 \cdot (-10) = $ _____."

15. Multiply.

    a. $(-10) \cdot 5 \cdot (-2)$

    b. $4 \cdot (-4) \cdot 0 \cdot (-9)$

    c. $100 \cdot (-1) \cdot (-3)$

    d. $(-3) \cdot (-2) \cdot (-5) \cdot (-2)$

    e. $2 \cdot (-5) \cdot (-10) \cdot 5 \cdot (-3)$

    f. $2 \cdot (-3) \cdot 4 \cdot (5)$

16. Multiply. Notice something interesting!

    $(-1) \cdot (-1) = $ _____

    $(-1) \cdot (-1) \cdot (-1) \cdot (-1) \cdot (-1) = $ _____

    $(-1) \cdot (-1) \cdot (-1) = $ _____

    $(-1) \cdot (-1) \cdot (-1) \cdot (-1) \cdot (-1) \cdot (-1) = $ _____

    $(-1) \cdot (-1) \cdot (-1) \cdot (-1) = $ _____

    $(-1) \cdot (-1) \cdot (-1) \cdot (-1) \cdot (-1) \cdot (-1) \cdot (-1) = $ _____

17. Find the values of the powers.

    a. $(-1)^6$

    b. $(-1)^9$

    c. $(-2)^3$

    d. $(-10)^4$

Puzzle Corner Solve. **a.** $(-1)^{101}$ **b.** $(-2)^7$ **c.** $(-10)^9$

Dividing Integers

Divide a negative number by a positive

The image illustrates $(-8) \div 4$, or eight negatives divided into four groups. We can see the answer is -2.

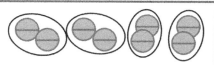

Any time a negative integer is divided by a positive integer, we can illustrate it as so many negative counters divided evenly into groups. The answer will be negative.

Divide a positive integer by a negative. For example, $24 \div (-8) = ?$

Remember, multiplication is the opposite operation to division. Let's write the answer to $24 \div (-8)$ as s. Then from that we can write a multiplication:

$24 \div (-8) = s \quad \Rightarrow \quad (-8)s = 24$

(You could use an empty line instead of s, if the variable s confuses you.)

The only number that fulfills the equation $(-8)s = 24$ is $s = -3$. Therefore, $24 \div (-8) = -3$.

Similarly, each time you divide a positive integer by a negative integer, the answer is <u>negative</u>.

Divide a negative integer by a negative. For example, $(-24) \div (-8) = ?$

Again, let's denote the answer to $-24 \div (-8)$ with y, and then write a multiplication sentence.

$-24 \div (-8) = y \quad \Rightarrow \quad (-8)y = -24$

The only number that fulfills the equation $(-8)y = -24$ is $y = 3$. Therefore, $-24 \div (-8) = 3$.

Similarly, each time you divide a negative integer by a negative integer, the answer is <u>positive</u>.

Summary. The symbols below show whether you get a positive or negative answer when you multiply or divide integers. Notice that the <u>rules for multiplication and division are the same</u>!

Multiplication	Examples	Division	Examples
⊕ · ⊖ = ⊖	$4 \cdot (-5) = -20$	⊕ ÷ ⊖ = ⊖	$20 \div (-5) = -4$
⊖ · ⊕ = ⊖	$-4 \cdot 5 = -20$	⊖ ÷ ⊕ = ⊖	$-20 \div 5 = -4$
⊖ · ⊖ = ⊕	$-4 \cdot (-5) = 20$	⊖ ÷ ⊖ = ⊕	$-20 \div (-5) = 4$
⊕ · ⊕ = ⊕	$4 \cdot 5 = 20$	⊕ ÷ ⊕ = ⊕	$20 \div 5 = 4$

Here is a shortcut for *multiplication* and *division* (NOT for addition or subtraction):

- If both numbers have the same sign (both are positive *or* negative), the answer is positive.
- If the numbers have different signs, the answer is negative.

1. Divide.

a. $-50 \div (-5) =$ _____	**b.** $(-8) \div (-1) =$ _____	**c.** $81 \div (-9) =$ _____
$-12 \div 2 =$ _____	$14 \div (-2) =$ _____	$-100 \div (-10) =$ _____

2. Multiply. Then use the same numbers to write an equivalent division equation.

a. $-5 \cdot (-5) = $ _____	b. $9 \cdot (-6) = $ _____	c. $-80 \cdot 8 = $ _____
_____ ÷ _____ = _____	_____ ÷ _____ = _____	_____ ÷ _____ = _____

3. Four people shared a debt of $280 equally. How much did each owe? Write an integer division.

4. In a math game, you get a negative point for every wrong answer and a positive point for every correct answer. Additionally, if you answer in 1 second, your negative points from the past get slashed in half!

   Angie had accumulated 14 negative and 25 positive points in the game. Then she answered a question correctly in 1 second. Write an equation for her current "point balance."

5. Complete the patterns.

a.	b.	c.
$12 \div 4 = $ _____	_____ $\div (-7) = -3$	$60 \div $ _____ $= 2$
$8 \div 4 = $ _____	_____ $\div (-7) = -2$	$40 \div $ _____ $= 2$
$4 \div 4 = $ _____	_____ $\div (-7) = -1$	$20 \div $ _____ $= 2$
$0 \div 4 = $ _____	_____ $\div (-7) = 0$	$-20 \div $ _____ $= 2$
$(-4) \div 4 = $ _____	_____ $\div (-7) = 1$	$-40 \div $ _____ $= 2$
$(-8) \div 4 = $ _____	_____ $\div (-7) = 2$	$-60 \div $ _____ $= 2$
$(-12) \div 4 = $ _____	_____ $\div (-7) = 3$	$-80 \div $ _____ $= 2$
$(-16) \div 4 = $ _____	_____ $\div (-7) = 4$	$-100 \div $ _____ $= 2$

6. Here's a funny riddle. Solve the math problems to uncover the answer.

   E _____ $\div (-8) = 2$ N $-12 \cdot (-5) = $ _____   E $(-144) \div 12 = $ _____

   E $3 \cdot (-12) = $ _____   H _____ $\div 12 = -5$ T $-4 \cdot (-9) = $ _____

   N $-15 \div $ _____ $= -5$ E _____ $\cdot (-6) = 0$ V $-45 \div $ _____ $= 5$

   G $-1 \cdot (-9) = $ _____   I $-27 \div 9 = $ _____   I $-7 \cdot $ _____ $= -84$

   S $-48 \div 6 = $ _____   N $3 \cdot $ _____ $= -24$

   Why is six afraid of seven? Because...

   -8 -12 -9 -36 60 0 12 9 -60 36 3 -3 -8 -16

Negative Fractions

When dividing negative integers, you may sometimes get a *negative fraction* as an answer.

Example 1. The division −6 ÷ 7 can be written as $\frac{-6}{7}$. Since we are dividing a negative integer by a positive one, the answer will be negative. The answer is the negative fraction $-\frac{6}{7}$. (Read: "negative six sevenths".)

The division 6 ÷ (−7) *also* gives us the same answer: $\frac{6}{-7}$ is equal to $-\frac{6}{7}$.

As a shortcut, think of **moving the minus sign from the numerator or the denominator to the front of the fraction**.

What do you think (−6) ÷ (−7) equals? Think for a bit!

Check the answer at the bottom of the page before you go on.

1. Divide, and mark the fractions on the number line. Remember to move the negative sign to the front of the fraction if your answer is a negative fraction. The first one is done for you.

a. $5 \div (-6) = \frac{5}{-6} = -\frac{5}{6}$	**b.** $-7 \div 6 =$	**c.** $-2 \div (-6) =$
d. $-13 \div 6 =$	**e.** $-10 \div (-6) =$	**f.** $21 \div (-6) =$

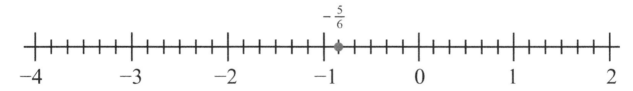

g. $-9 \div 8 =$	**h.** $16 \div (-8) =$	**i.** $-17 \div (-8) =$
j. $-25 \div 8 =$	**k.** $-5 \div (-4) =$	**l.** $8 \div (-4) =$

The answer to $(-6) \div (-7)$ or $\frac{-6}{-7}$ is $\frac{6}{7}$. Dividing a negative by a negative gives us a positive.

2. Divide and simplify.

a. $2 \div (-4) = -\dfrac{2}{4} = -\dfrac{1}{2}$	b. $4 \div (-10)$	c. $-10 \div 4$
d. $-1 \div (-5)$	e. $15 \div (-25)$	f. $-12 \div (-8)$
g. $-56 \div 72$	h. $-40 \div (-12)$	i. $45 \div (-55)$

There is one kind of division you cannot do, and that is **a division by zero**. $\dfrac{-6}{0}$ is not allowed!

"Why?" you might ask. Couldn't the answer be zero?

Let's see what happens if $-6 \div 0 = 0$. Then the corresponding multiplication would be $0 \cdot 0 = 6$. That is not correct. So if we were to make the answer 0, it would break other rules of math. Since $x \cdot 0 = 6$ has no solution, mathematicians simply say that $-6 \div 0 = x$ is undefined.

3. **a.** Find the value of the expression $5/x$ for different values of the variable x.

x	$\dfrac{5}{x}$	x	$\dfrac{5}{x}$	x	$\dfrac{5}{x}$
4		1		−2	
3		0		−3	
2		−1		−4	

  b. Now think about this. Imagine you continued the table and included even more values of x and $5/x$.
  Will you always get a fraction if x is not zero?
  In other words, is $5/x$ always a fraction, assuming $x \neq 0$?

4. **a.** Choose values for m and n so that the quotient m/n is a negative whole number.

  b. Choose values for m and n so that $m < 0$ and the quotient m/n is positive.

5. These are the maximum temperatures for a week in January in Ryan's hometown.
   Find the average maximum temperature.

   −5°C, −2°C, 1°C, 2°C, −7°C, −5°C, −2°C

A neat little trick for calculating averages

Example 2. Calculate the average weight of kittens that weigh 98 g, 101 g, 96 g, 105 g, and 99 g.

All the weights are near 100 g. Let's *guess* the average is 100 g. Now, instead of calculating with the actual weights, consider their *differences* from 100 g. For a weight that is less than 100 g, the difference will be negative.

The differences are: −2 g, 1 g, −4 g, 5 g, and −1 g. Now, calculate the average of the <u>differences</u>:

$$\frac{-2\text{ g} + 1\text{ g} + (-4\text{ g}) + 5\text{ g} + (-1\text{ g})}{5} = \frac{-1\text{ g}}{5} = -\frac{1}{5}\text{ g, or as a decimal, } -0.2\text{ g.}$$

We get the true average by adding the difference we calculated to the average we guessed:
100 g + (−0.2 g) = 99.8 g. Yes, it was indeed close to 100 g!

---

6. Look at the graph below.

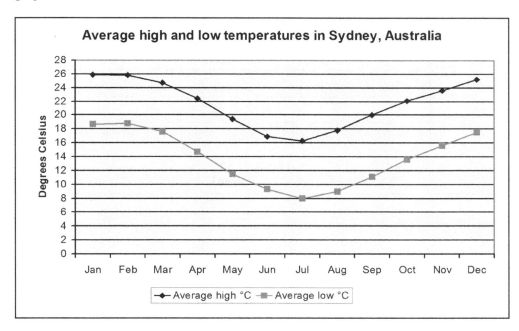

Let's guess that the average high temperature over the whole year in Sydney is 20°C. Read the monthly high temperatures from the graph the best you can, and then calculate the average high temperature by using the method above.

If you need more practice, repeat this for the average low temperatures.

7. The chart shows Alice's earnings and expenditures on a weekly basis.

	Income	Expenses	Surplus/Deficit
Week 12	$12	$10	$2
Week 13	$15	$12	$3
Week 14	$8	$10	−$2
Week 15	$18	$9	
Week 16	$5	$18	
Week 17	$8	$13	
Week 18	$14	$9	

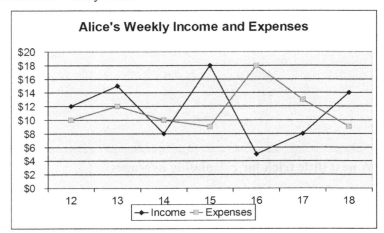

From this data, she made *another* graph that only shows whether she had a surplus or a deficit in any particular week, in other words, whether she earned or lost money overall. For example, in week 12, she had a surplus or an excess of $2, because $12 − $10 = $2.

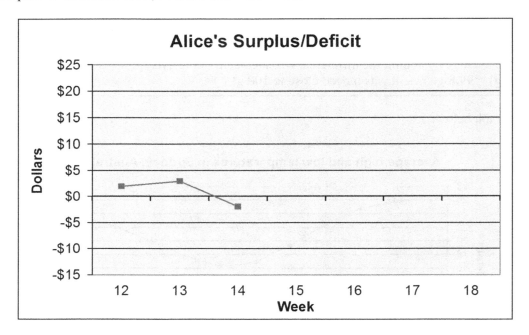

a. Finish drawing the line graph.

b. Calculate the average surplus/deficit over the 7-week period.

c. Is it negative or positive?

d. What does that mean?

The Order of Operations

You are used to performing more than one operation at a time with positive numbers, such as $12 \cdot 5 \cdot 7$ or $120 \div 4 + 30$. We can also have several operations with integers in one expression. **The order of operations is still: parentheses, exponentiation, multiplication and division, and addition and subtraction (PEMDAS).**

Example 1. Find the value of $30 - 5 \cdot (-7)$.
There is a subtraction and a multiplication. According to the order of operations, the multiplication is done first. To help you see it, you can put brackets around it:

$$30 - 5 \cdot (-7)$$
$$= 30 - [5 \cdot (-7)]$$

We solve $5 \cdot (-7) = -35$ first. The expression becomes

$30 - (-35) = 30 + 35 = 65$.

Example 2. $-5 \cdot 8 - \dfrac{20}{-4}$

Here we have a multiplication, a subtraction, and a division. The multiplication and division are done first, before the subtraction. We get:

$$-5 \cdot 8 - \dfrac{20}{-4}$$
$$\downarrow \qquad \downarrow$$
$$-40 - (-5) = -40 + 5 = -35$$

1. Find the value of the expressions using the correct order of operations.

a. $-8 + 5 \cdot (-3)$	b. $5 \cdot (2 - 9)$	c. $-5 \cdot 7 - 4 \cdot 3$
d. $4 + (-5) \cdot 5 - 6$	e. $3 - 5 \cdot (-5)$	f. $(-5 + 4) \cdot 8$
g. $8 \cdot \dfrac{-8}{4}$	h. $1 - \dfrac{2}{-5}$	i. $-8 + \dfrac{21}{4 - 7}$

2. Find the value of the expressions when $x = -2$ and $y = 5$.

a. $x(y + 1)$	b. $xy + 1$	c. $-2(x + y)$

3. Draw lines to connect the expressions that have the same *value*.

$-5 \cdot 1 + 5$	$3 \cdot 2$	$2 \cdot (-2) - 2$
$-7 + (-2 + 3)$	$10 + (-10)$	$(-7 - (-2) + 5) \cdot 3$
$(-2) \cdot (-3)$	$-2 \cdot 3$	$-2 \cdot 2 + 10$

4. The points (−8, 6), (4, 4), (6, −6), and (−6, −2) are vertices of a quadrilateral.

   a. Draw the quadrilateral.

   b. Divide each coordinate of each point by 2, to get four new points. Write the coordinates of the new points:

   c. Draw a new quadrilateral using the four points from (b) as vertices.

   What you just did was to *reduce* the original quadrilateral. The original and the new quadrilateral are **similar quadrilaterals**: They have the same shape.

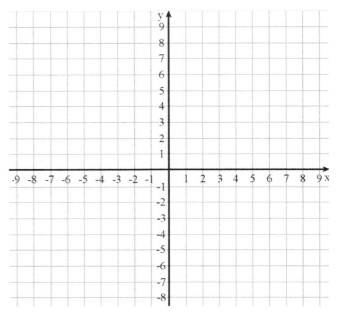

5. Add a minus sign in front of one number in each equation to make each equation true.
   Note: There may be more than one way to do this.

| **a.** $45 \div 5 = -9$ | **b.** $-8 \cdot 7 = 56$ | **c.** $60 \div (-5) = 12$ |

6. Write two different division equations for each quotient.

| **a.** ___ ÷ ___ = −7
 ___ ÷ ___ = −7 | **b.** ___ ÷ ___ = −1
 ___ ÷ ___ = −1 | **c.** ___ ÷ ___ = −5
 ___ ÷ ___ = −5 |

7. Solve the equations by thinking logically.

| **a.** $-10y = 100$
 $y = $ ___ | **b.** $4a = -36$
 $a = $ ___ | **c.** $w \div 3 = -6$
 $w = $ ___ |
| **d.** $\dfrac{b}{-5} = 15$
 $b = $ ___ | **e.** $\dfrac{-55}{5} = x$
 $x = $ ___ | **f.** $\dfrac{-64}{z} = 8$
 $z = $ ___ |

Puzzle Corner

If you multiply both coordinates of the point (−2, 3) by −1, it becomes (2, −3). Investigate what happens to a geometric figure if you multiply each of its coordinates by −1.

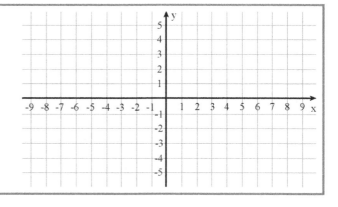

Chapter 2 Mixed Review

1. Find the value of the expressions.

   a. $(3 \text{ cm})^3$ **b.** $(1.5 \text{ in})^2$ **c.** $(10 \text{ m})^3$

2. Write an expression for the volume of a cube with edge 2*s*.

3. Evaluate the expression from exercise 2 when *s* = 1.7 cm.

4. Write an equation. Then solve it using mental math.

   a. 201 decreased by a number is 167.

   b. The product of a number and 7 equals 7/24.

Equation	Solution

5. Is subtraction commutative? Explain your reasoning using an example.

6. Name the property of arithmetic illustrated by the equation $(n + 2) + 5 = n + (2 + 5)$.

7. Fill in the table.

Expression	Terms in it	Coefficient(s)	Constant(s)
$(5/6)s^2$			
$x + 2y + 8$			
$p \cdot 46$			

8. Write an expression for the area in two ways, thinking of the overall rectangle or its component rectangles.

____ (____ + ____) and

3 · ____ + ____ · ____

9. Use the distributive property "backwards" to write the expression as a product (it is called factoring).

a. $3x + 6 =$ ____ (____ + ____)

b. $10z - 20 =$ ____ (____ − ____)

10. Write using symbols, and simplify if possible.

   a. The opposite of the absolute value of 2

   b. The absolute value of the opposite of 2

11. Write an inequality. Use negative integers where appropriate.

   a. The ditch is at least 2 ft deep.

   b. Ashley owes me no more than $100.

   c. The freezer temperature shouldn't be colder than 20 Celsius degrees below zero.

12. The fence that surrounds Emily's yard is x feet long. Emily has painted 1/4 of it. Write an expression for the length of the fence that is *not* painted.

13. Sketch a rectangle with an area of $3x + 21$.

14. Simplify the expressions.

a. $t + 2t + 9 - t$	**b.** $x \cdot 4 \cdot x \cdot x$	**c.** $5 \cdot y \cdot y \cdot x \cdot 2$

Chapter 2 Review

1. Match the equations with the situations and complete the missing parts.

    a. A ball was dropped from 18 ft above sea level; it fell 12 ft.
       Now the ball is at _____ ft.

    b. John had a $12 debt. He earned $18. Now he has _____.

    c. John had $12. He had to pay his dad $18. Now he has _____.

    d. A diver was at the depth of 18 ft. Then he rose 12 ft.
       Now he is at _____ ft.

    e. The temperature was −12°C and fell 18°. Now it is _____ °C.

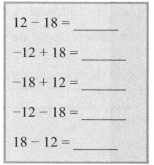

2. Compare the integers using > or < . Include the units, too.

    a. The temperature inside the fridge is 5°C.
       In the freezer, it is 12°C below zero.

    b. Andy has a debt of $400. Harry owes $250.

3. Add or subtract.

| a. (−12) + (−1) = _____ | b. −12 − (−1) = _____ | c. 7 − 12 = _____ |

4. Complete the equations, using <u>one positive</u> and <u>one negative</u> integer. There are many possible solutions.

| a. ____ + ____ = −2
 ____ + ____ = −2 | b. ____ + ____ = 0
 ____ + ____ = 0 | c. ____ + ____ = 3
 ____ + ____ = 3 |

5. Interpret the absolute value in each situation.

    a. A shark is swimming at the depth of 14 m. $|-14 \text{ m}|$ = _____ m

       Here, the absolute value shows _____

    b. Shelley's balance is −$31. $|-\$31|$ = $ _____

       Here, the absolute value shows _____

6. Write using symbols, and simplify if possible.

    a. the opposite of 8

    b. the opposite of −100

    c. the opposite of the sum 2 + 5

    d. the absolute value of negative 45

7. Write an addition or subtraction sentence to match the temperature change.

   a. The temperature was −2°C. Then it rose 4 degrees. Now it is _____.

   b. The temperature was −11°C. Then it rose 5 degrees. Now it is _____.

   c. The temperature was 2°C. It dropped 8 degrees. Now it is _____.

   d. The temperature was −3°C. It dropped 8 degrees. Now it is _____.

   ⁻2 + 4 = ____

8. Plot these inequalities on the number line.

   a. $x \geq -2$

   b. $x < 2$

9. Add.

   a. $(-2) + 7 + (-7) + (-1) =$ _____

   b. $4 + (-10) + (-12) + 1 =$ _____

10. Iodide is an ion with 53 protons and 54 electrons. What is the total electric charge of this ion?

11. The top of a fishing net is at a depth of 6 m below the surface of a lake, and later it is lowered to the depth of 8 m. Write an expression for the distance between these two depths using negative integers.

12. Solve.

    a. $21 + (-48) =$

    b. $41 + (-38) =$

    c. $-610 + 900 =$

13. Change each subtraction into an addition and solve.

    a. $1 - (-7)$
    ↓
    ____ + ____ = ____

    b. $2 - (-11)$
    ↓
    ____ + ____ = ____

    c. $-20 - (-6)$
    ↓
    ____ + ____ = ____

    d. $3 - 8$
    ↓
    ____ + ____ = ____

14. Which king ruled the Persian Empire longer, Xerxes I, who ruled from 486 to 465 BC, or Darius II, who ruled from 424 to 404 BC?

15. Multiply.

    a. $-2 \cdot (-4) =$ _____

    $-2 \cdot 4 =$ _____

    b. $(-3) \cdot (-8) =$ _____

    $7 \cdot (-12) =$ _____

    c. $(-3) \cdot 3 \cdot (-1) =$ _____

    $-7 \cdot (-2) \cdot (-2) =$ _____

16. True or false?

   a. Any integer more than 6 has an absolute value more than 6.

   b. Any integer less than 6 has an absolute value less than 6.

   c. A number and its opposite have the same absolute value.

   d. The absolute value of the opposite of a number is the same as the opposite of the absolute value of the same number.

17. Divide.

a. $-10 \div (-5) = $ _____	b. $(-12) \div (-4) = $ _____	c. $-56 \div 7 = $ _____
$24 \div (-3) = $ _____	$21 \div (-3) = $ _____	$-120 \div (-10) = $ _____

18. Find the missing numbers.

a. $-5 \cdot $ _____ $= -30$	b. $2 \cdot $ _____ $= -18$	c. $-8 \cdot $ _____ $= 48$
d. $-42 \div $ _____ $= 6$	e. $-64 \div $ _____ $= -8$	f. $81 \div $ _____ $= -9$

19. Solve the equations by thinking of multiplication tables.

a. $5y = -100$	b. $-4b = -48$	c. $\dfrac{35}{-5} = y$
$y = $ _____	$b = $ _____	$y = $ _____

20. Give a real-life situation for the product $3 \cdot (-10)$.

21. Divide and simplify if possible.

a. $1 \div (-6)$	b. $-3 \div 15$	c. $-6 \div (-7)$

22. Find the value of the expressions when $x = -3$ and $y = 4$.

a. x^2	b. $-5xy$	c. $2 - (y + x)$

Chapter 3: One-Step Equations
Introduction

The goal of this chapter is to solve one-step equations that involve integers. The first lesson reviews the concept of an equation and how to model equations using a pan balance (scale). Recall that the basic principle for solving equations is that, when you perform the same operation on both sides of an equation, the two sides remain equal.

The chapter presents two alternatives for keeping track of the operations to be performed on an equation. The one method, writing the operation under each side of the equation, is common in the United States. The other method, writing the operation in the right margin, is common in Finland. Either is adequate, and the choice is just a matter of the personal preference of the teacher.

The introduction to solving equations is followed by a lesson on addition and subtraction equations and another on multiplication and division equations. All the equations are easily solved in only one step of calculations. The twofold goal is to make the student proficient in manipulating negative integers and also to lay a foundation for handling more involved equations in Chapter 5.

In the next lesson, students write equations to solve simple word problems. Even though they could solve most of these problems without using the equations, the purpose of the lesson is to make the student proficient in writing simple equations before moving on to more complex equations from more difficult word problems.

The last topic, in the lesson *Constant Speed*, is solving problems with distance (d), rate or velocity (v), and time (t). Students use the equivalent formulas $d = vt$ and $v = d/t$ to solve problems involving constant or average speed. They learn an easy way to remember the formula $v = d/t$ from the unit for speed that they already know, "miles per hour."

As a reminder, it is not recommended that you assign all the exercises by default. Use your judgment, and strive to vary the number of assigned exercises according to the student's needs. Some students might only need half or even less of the available exercises, in order to understand the concepts.

The Lessons in Chapter 3

	page	span
Solving Equations	82	*7 pages*
Addition and Subtraction Equations	89	*4 pages*
Multiplication and Division Equations	93	*4 pages*
Word Problems	97	*3 pages*
Constant Speed	100	*7 pages*
Chapter 3 Mixed Review	107	*3 pages*
Chapter 3 Review	110	*2 pages*

Helpful Resources on the Internet

You can also access this list of links at **https://links.mathmammoth.com/gr7ch3**

The Simplest Equations - Video Lessons by Maria
A set of free videos that teach the topics in this chapter - by the author.
https://www.mathmammoth.com/videos/prealgebra/pre-algebra-videos.php#equations

Solving Equations by Balancing
Practice solving for *x* while keeping the equation balanced in this interactive online activity.
https://www.geogebra.org/m/M8rYtDF9

Balance When Adding and Subtracting
Click on the buttons above the scales to add or subtract until you can figure out the value of *x* in the equation.
https://www.mathsisfun.com/algebra/add-subtract-balance.html

Addition and Subtraction Equations
Use this interactive balance model to practice solving for *x*.
https://www.geogebra.org/m/xfqjck9q

Solving Linear Equations Quiz
Practice solving simple one-step linear equations using properties of multiplication and division.
https://www.softschools.com/quizzes/math/solving_linear_equations/quiz2013.html

Modeling with One-Step Equations
Practice writing basic equations to model real-world situations in this interactive activity from Khan Academy.
https://bit.ly/Modeling-with-One-Step-Equations

Equation Balance Model
Balance the equation represented on the scale by adding the same number to both sides. The equations given are one-step addition and subtraction equations.
https://www.geogebra.org/m/xfqjck9q

Equation Pal - Online
This program allows students to solve equations on a step-by-step basis in an "interview" format.
https://mrnussbaum.com/equation-pal-online

Algebra Four
A connect four game with equations. For this level, choose difficulty "Level 1" and "One-Step Problems".
http://www.shodor.org/interactivate/activities/AlgebraFour/

One-Step Equation Game
Choose the correct root for the given equation (multiple-choice), and then you get to attempt to shoot a basket.
https://www.math-play.com/One-Step-Equations-Basketball/One-Step-Equations-Basketball_html5.html

Arithmagons
Find the numbers that are represented by question marks in this interactive puzzle.
https://www.transum.org/Software/SW/Starter_of_the_day/starter_August20.ASP

Cars
Use clues to help you find the total cost of four cars in this fun brainteaser.
https://www.transum.org/Software/SW/Starter_of_the_day/starter_July16.ASP

Distance, Speed, and Time from BBC Bitesize
Instruction, worked out exercises, and an interactive quiz relating to constant speed, time, and distance. A triangle with letters D, S, and T helps students remember the formulas for distance, speed, and time.
https://www.bbc.co.uk/bitesize/guides/z4swxnb/revision/1

Speed and Velocity
A short illustrated lesson about speed and velocity. There is a quiz at the end of the page, which includes both easier and challenging questions.
https://www.mathsisfun.com/measure/speed-velocity.html

Solving Equations

Do you remember? An **equation** has two expressions, separated by an equal sign:

$$(\text{expression}) = (\text{expression})$$

To solve an equation, we can

- add the same quantity to both sides
- subtract the same quantity from both sides
- multiply both sides by the same number
- divide both sides by the same number

Notice that in any of these operations, the two expressions on the left and right sides of the equation will remain equal, even though the expressions themselves change!

Example 1. We will manipulate the simple equation $2 + 3 = 5$ in these four ways. We will write in the margin the operation that is going to be done next to both sides.

Let's add six to both sides.	$2 + 3 = 5$	$\vert\, + 6$
Now, both sides equal 11. Next, we multiply both sides by 8.	$2 + 3 + 6 = 11$	$\vert\, \cdot 8$
Now, both sides equal 88. Next, we subtract 12 from both sides.	$16 + 24 + 48 = 88$	$\vert\, - 12$
Now both sides equal 76. Next, we divide both sides by 2.	$16 + 24 + 48 - 12 = 76$	$\vert\, \div 2$
Now both sides equal 38.	$8 + 12 + 24 - 6 = 38$	

Of course, you do not usually work with equations like the one above, but with ones that have an unknown. Your goal is to **isolate** the unknown, or **leave it by itself,** on one side. Then the equation is solved.

We can model an equation with a **pan balance**. Both sides (pans) of the balance will have an *equal* weight in them, thus the sides are balanced (not tipped to either side).

Example 2. Solve the equation $x - 2 = 3$.

We can write this equation as $x + (-2) = 3$ and model it using negative and positive counters in the balance.

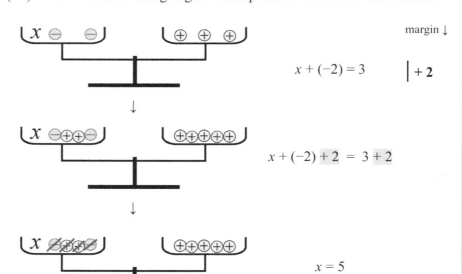

Here x is accompanied by two negatives on the left side. Adding two <u>positives</u> *to both sides* will cancel those two negatives. We denote that by writing "**+2**" in the margin.

We write $x + (-2) + 2 = 3 + 2$ to show that 2 was added to both sides of the equation.

Now the two positives and two negatives on the left side cancel each other, and x is left by itself. On the right side we have 5, so x equals 5 positives.

1. Solve the equations. Write in the margin what operation you do to both sides.

a. Balance	Equation	Operation to do to both sides
	$x + 1 = -4$	

b. Balance	Equation	Operation to do to both sides
	$x - 1 = -3$	

c. Balance	Equation	Operation
	$x - 2 = 6$	

d. Balance	Equation	Operation
	$x + 5 = 2$	

2. If you need more practice, solve the following equations also. Draw a balance in your notebook to help you.

   a. $x + (-3) = 7$ **b.** $x + (-3) = -4$ **c.** $x + 6 = -1$ **d.** $x + 5 = -4$

In the two examples below, we either multiply or divide both sides by the same number. Study them carefully!

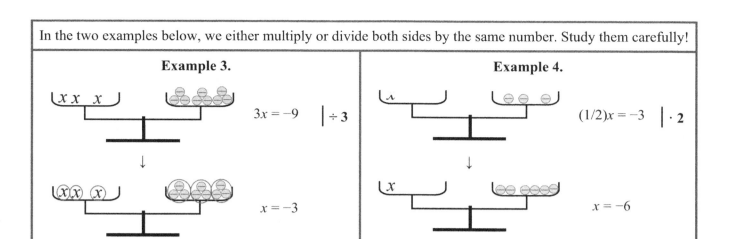

Example 3.

$3x = -9 \quad | \div 3$

$x = -3$

Example 4.

$(1/2)x = -3 \quad | \cdot 2$

$x = -6$

3. Solve the equations. Write in the margin what operation you do to both sides.

a. Balance	Equation	Operation to do to both sides
	$4x = -12$	

b. Balance	Equation	Operation to do to both sides
	$(1/3)x = -5$	

4. Let's review a little! Which equation matches the situation?

   a. A stuffed lion costs $8 less than a stuffed elephant.
      Note: p_l signifies the price of the lion and p_e the price of the elephant.

   $\boxed{p_e = 8 - p_l}$ $\boxed{p_e = p_l - 8}$ $\boxed{p_l = p_e - 8}$

   b. A shirt is discounted by 1/5, and now it costs $16.

   $\boxed{p - 1/5 = \$16}$ $\boxed{\dfrac{4p}{5} = \$16}$ $\boxed{\dfrac{p}{5} = \$16}$ $\boxed{\dfrac{5p}{4} = \$16}$ $\boxed{\dfrac{p}{5} = 4 \cdot \$16}$

5. Find the roots of the equation $\dfrac{6}{x+1} = -3$ in the set $\{2, -2, 3, -3, 4, -4\}$.

6. Write an equation, then solve it using guess and check. Each root is between −20 and 20.

a. 7 less than x equals 5.

b. 5 minus 8 equals x plus 1

c. The quantity x minus 1 divided by 2 is equal to 4.

d. x cubed equals 8

e. −3 is equal to the quotient of 15 and y

f. Five times the quantity x plus 1 equals 10.

Example 5. Solve $2x + 5 = -3$. The solution requires two steps.

The two *x*'s are accompanied by five positives. Therefore, we will subtract five *from both sides*.

Subtracting 5 is the same as adding −5, so the right side ends up with 8 negatives.

The positives and negatives on the left side cancel each other, and $2x$ is left by itself on that side.

Now we need to divide both sides by 2. Again, we note that in the margin.

We can see that *x* equals 4 negatives.

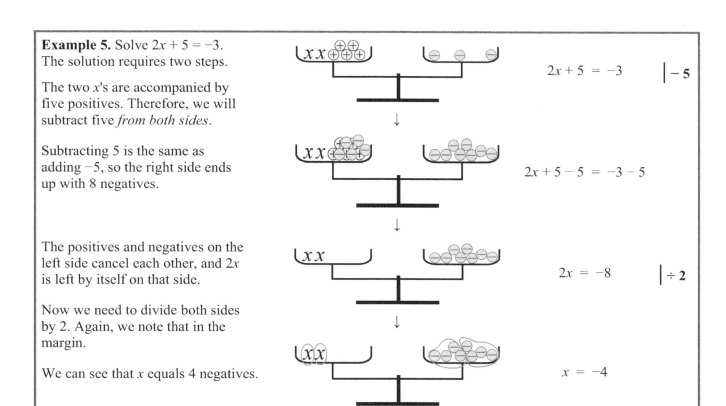

$2x + 5 = -3 \quad | -5$

$2x + 5 - 5 = -3 - 5$

$2x = -8 \quad | \div 2$

$x = -4$

7. Solve the equations. Write in the margin what operation you do to both sides.

a. Balance	Equation	Operation to do to both sides
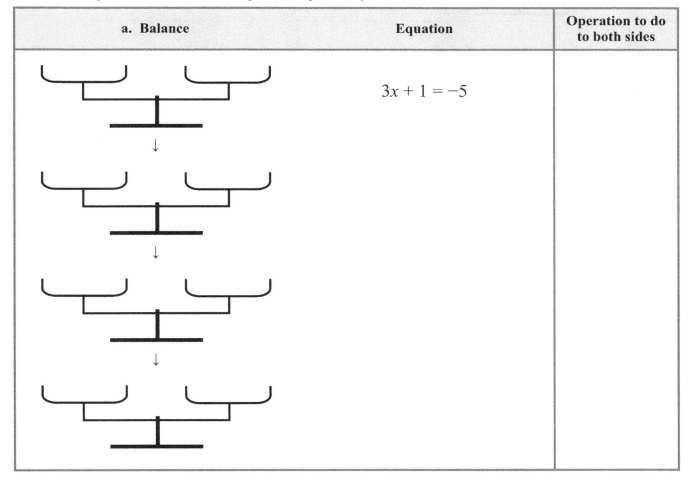	$3x + 1 = -5$	

b. Balance	Equation	Operation
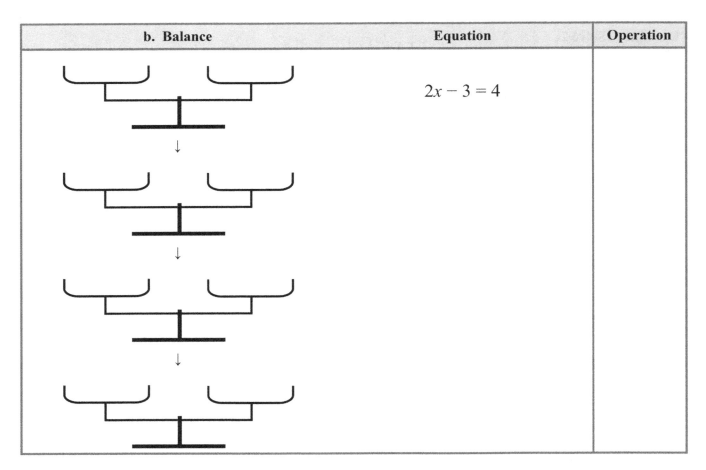	$2x - 3 = 4$	

c. Balance	Equation	Operation
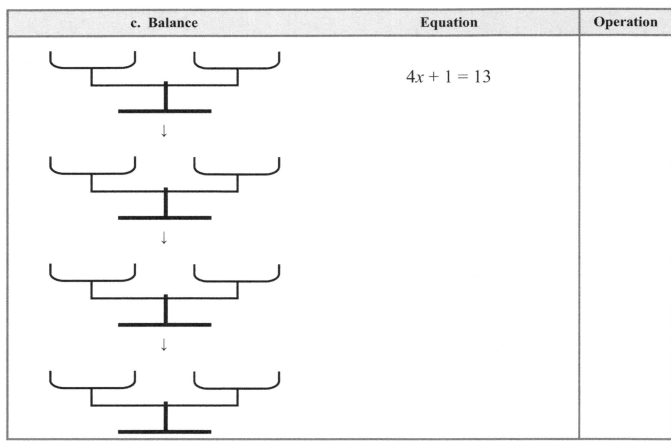	$4x + 1 = 13$	

Addition and Subtraction Equations

You can **keep track of the operations you're using in a couple of different ways**. One way is to write the operation underneath the equation on both sides. Another is to write it in the right margin, like we did in the last lesson.	**One way:** $x + 9 = 4$ $\underline{-9 \quad -9}$ $x = -5$	**Another way:** $x + 9 = 4 \quad \vert -9$ $x + 9 - 9 = 4 - 9$ (This step is optional.) $x = -5$
But in either case, always check your solution: does it solve the original equation?	Check: $-5 + 9 \stackrel{?}{=} 4$. Yes, it checks.	

1. Solve these one-step equations. Keep track of the operations either under the equation or in the margin, whichever way your teacher prefers.

a. $\quad x + 5 = 9$	b. $\quad x + 5 = -9$
c. $\quad x - 2 = 3$	d. $\quad w - 2 = -3$
e. $\quad z + 5 = 0$	f. $\quad y - 8 = -7$

2. In these equations, your first step is to <u>simplify</u> what is on the right side.

a. $\quad x - 7 = 2 + 8$ $\quad x - 7 = 10$	b. $\quad x - 10 = -9 + 5$
c. $\quad s + 5 = 3 + (-9)$	d. $\quad t + 6 = -3 - 5$

If the **unknown is on the right side of the equation**, you have two options:	1. Flip the sides first:	2. Solve as it is:
• First, flip the two sides. Then solve as usual. • Or, solve as usual, isolating the unknown —this time on the right side of the equation. The solution will initially read as $-7 = x$. Flip the sides now and write the solution as $x = -7$.	$-9 = x - 2$ $x - 2 = -9$ $\underline{+2 \quad\quad +2}$ $x \quad = \quad -7$	$-9 = x - 2$ $\underline{+2 \quad\quad +2}$ $-7 = x$ $x = -7$

3. Solve. Check your solutions.

a.	$-8 = s + 6$	b.	$-2 = x - 7$
c.	$4 = s + (-5)$	d.	$2 - 8 = y + 6$
e.	$5 + x = -9$	f.	$-6 - 5 = 1 + z$
g.	$y - (-7) = 1 - (-5)$	h.	$6 + (-2) = x - 2$
i.	$3 - (-9) = x + 5$	j.	$2 - 8 = 2 + w$

Example 1. Solve $-2 + 8 = -x$.

Our first step is to simplify the sum $-2 + 8$.

The equation becomes $-x = 6$ (or $6 = -x$). What does that mean? It means that the opposite of x is 6. So x must equal -6!

Lastly we check the solution $x = -6$:

$-2 + 8 \stackrel{?}{=} -(-6)$, which simplifies to $6 \stackrel{?}{=} 6$, so it checks.

1. Flip the sides first:	2. Solve as it is:
$-2 + 8 = -x$	$-2 + 8 = -x$
$-x = -2 + 8$	$6 = -x$
$-x = 6$	$-6 = x$
$x = -6$	$x = -6$

4. Solve for x. Check your solutions.

a.	$-x = 6$	**b.**	$-x = 5 - 9$
c.	$4 + 3 = -y$	**d.**	$-2 - 6 = -z$

5. Which equation best matches the situation?

   a. The sides of a square playground were shortened by 1/2 m, and now its perimeter is 12 m.

   $4s - 1/2 = 12$ $4(s - 1/2) = 12$ $4s - 50 = 12$

   $s - 1/2 = 4 \cdot 12$ $s - 1/2 = 12$ $4(s - 0.5) = 12$

   b. How long were the sides before they were made shorter? Solve the problem using mental math.

   c. *Challenge*: Solve the same problem using the equation. Compare the steps of this formal solution to the way you reasoned it out in your head. Are the steps similar?

6. Here is another "growing pattern." Draw steps 4 and 5 and answer the questions.

Step   1   2   3

   a. How do you see this pattern grow?

   b. How many flowers will there be in step 39?

   c. In step n?

Example 2. Solve $8 - x = -2$.

As usual, think about what you need to do to isolate the x on one side. Since there is an 8 on the side with the x, we need to subtract 8 from both sides.

However, note that x is being *subtracted*, or in other words, there is a negative sign in front of the x. This negative sign <u>does not disappear</u> when you subtract 8 from both sides.

Writing the operation underneath each side:

$$\begin{aligned} 8 - x &= -2 \\ -8 & \quad -8 \\ \hline -x &= -10 \\ x &= 10 \end{aligned}$$

Writing the operation in the right margin:

$$\begin{aligned} 8 - x &= -2 \quad |-8 \\ 8 - x - 8 &= -2 - 8 \\ -x &= -10 \\ x &= 10 \end{aligned}$$

If this is confusing, think of it this way: The equation $8 - x = -2$ can also be written as $8 + (-x) = -2$. When we subtract 8 from both sides, the left side becomes $8 + (-x) - 8$. The positive 8 and negative 8 will cancel each other and leave $-x$.

So we end up with the equation $-x = -10$. This equation says that the opposite of x is negative 10, so x must be 10. (Why?)

Lastly, check your solution by substituting $x = 10$ back into the original equation: $8 - \underline{10} \stackrel{?}{=} -2$ ✓

7. Solve. Check your solutions.

a.	$2 - x = 6$		**b.**	$8 - x = 7$
c.	$-5 - x = 5$		**d.**	$2 - x = -6$
e.	$1 = -5 - x$		**f.**	$2 + (-9) = 8 - z$
g.	$-8 + r = -5 + (-7)$		**h.**	$2 - (-5) = 2 + 5 + t$

Multiplication and Division Equations

Do you remember **how to show simplification**? Just cross out the numbers and write the new numerator above the fraction and the new denominator below it. Notice that the number you divide by (the 5 in the fraction at the right) isn't indicated in any way!	$\dfrac{\cancel{35}^{\,7}}{\cancel{55}_{\,11}} = \dfrac{7}{11}$
We can simplify expressions involving variables in exactly the same way. In the examples on the right, we cross out the *same number* from the numerator and the denominator. That is based on the fact that a number divided by itself is 1. We could write a little "1" beside each number that is crossed out, but that is usually omitted.	$\dfrac{\cancel{2}x}{\cancel{2}} = x \qquad \dfrac{\cancel{5}s}{\cancel{5}} = s$ $\dfrac{4\cancel{x}}{\cancel{x}} = 4$
In this example, we simplify the fraction 3/6 into 1/2 the usual way.	$\dfrac{\cancel{3}^{\,1}x}{\cancel{6}_{\,2}} = \dfrac{1}{2}x \text{ or } \dfrac{x}{2}$
Notice: We divide both the numerator and the denominator by 8, but <u>this leaves −1 in the denominator</u>. Therefore, the whole expression simplifies to −z instead of z.	$\dfrac{\cancel{8}z}{\cancel{-8}} = \dfrac{z}{-1} = -z$

1. Simplify.

a. $\dfrac{8x}{8}$		b. $\dfrac{8x}{2}$		c. $\dfrac{2x}{8}$	
d. $\dfrac{-6x}{-6}$		e. $\dfrac{-6x}{6}$		f. $\dfrac{6x}{-6}$	
g. $\dfrac{6w}{2}$		h. $\dfrac{6w}{w}$		i. $\dfrac{6w}{-2}$	

2. Draw the fourth and fifth steps of the pattern and answer the questions.

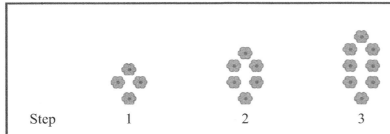

Step   1   2   3

a. How would you describe the growth of this pattern?

b. How many flowers will there be in step 39?

c. In step *n*?

Now you should be ready to use multiplication and division to solve simple equations.

Example 1. Solve $-2x = 68$.

The unknown is being multiplied by -2. To isolate it, we need to divide both sides by -2. (See the solution on the right.)

We get $x = -34$. Lastly we check the solution by substituting -34 in the place of x in the original equation:

$-2(-34) \stackrel{?}{=} 68$

$68 = 68$ It checks.

$-2x = 68$ This is the original equation.

$\dfrac{-2x}{-2} = \dfrac{68}{-2}$ We divide both sides by -2.

$\dfrac{\cancel{-2}x}{\cancel{-2}} = \dfrac{68}{-2}$ Now it is time to simplify. We cross out the -2 factors on the left side. On the right side, we do the division.

$x = -34$ This is the final answer.

Note: Most people combine the first 3 steps into one when writing the solution. Here they are written out for clarity.

3. Solve. Check your solutions.

a. $5x = -45$	b. $-3y = -21$
c. $-4 = 4s$	d. $72 = -6y$

4. Solve. Simplify the one side first.

a. $-5q = -40 - 5$	b. $2 \cdot 36 = -6y$
c. $3x = -4 + 3 + (-2)$	d. $5 \cdot (-4) = -10z$

Example 2. Solve $\dfrac{x}{-6} = -5$.

Here the unknown is divided by −6. To undo that division, we need to *multiply* both sides by −6. (See the solution on the right.)

We get $x = 30$. Lastly we check the solution:

$$\dfrac{30}{-6} \stackrel{?}{=} -5$$

$$-5 = -5 \checkmark$$

$\dfrac{x}{-6} = -5$ This is the original equation.

$\dfrac{x}{-6} \cdot (-6) = -5 \cdot (-6)$ We multiply both sides by −6.

$\dfrac{x}{\cancel{-6}} \cdot (\cancel{-6}) = 30$ Now it is time to simplify. We cross out the −6 factors on the left side, and multiply on the right.

$x = 30$ This is the final answer.

When writing the solution, most people would combine steps 2 and 3. Here both are written out for clarity.

5. Solve. Check your solutions.

a. $\dfrac{x}{2} = -45$

b. $\dfrac{s}{-7} = -11$

c. $\dfrac{c}{-7} = 4$

d. $\dfrac{a}{-13} = -9 + (-11)$

6. Write an equation for each situation. Then solve it. Do not write the answer only, as the main purpose of this exercise is to practice writing equations.

   a. A submarine was located at a depth of 500 ft. There was a shark swimming at 1/6 of that depth. At what depth is the shark?

   b. Three towns divided highway repair costs equally. Each town ended up paying $21,200. How much did the repairs cost in total?

Example 3. Solve $-\frac{1}{5}x = 2$. Here the unknown is multiplied by a negative fraction, but do not panic!

You see, you can *also* write this equation as $\frac{x}{-5} = 2$, where the unknown is simply divided by negative 5.

So what should we do in order to isolate *x*?

That is correct! Multiplying by −5 will isolate *x*. In the boxes below, this equation is solved in two slightly different ways, though both are doing essentially the same thing: multiplying both sides by −5.

Multiplying a fraction by its reciprocal:	**Canceling a common factor:**
$-\frac{1}{5}x = 2 \quad \mid \cdot (-5)$	$-\frac{1}{5}x = 2$ rewrite the equation
$(-5) \cdot \left(-\frac{1}{5}\right)x = (-5) \cdot 2$ Note that $-5 \cdot (-1/5)$ is 1.	$\frac{x}{-5} = 2 \quad \mid \cdot (-5)$
$1x = -10$	$\frac{x}{-5} \cdot (-5) = 2 \cdot (-5)$
$x = -10$	$x = -10$

Lastly we check the solution by substituting −10 in place of *x* in the original equation:

$$-\frac{1}{5}(-10) \stackrel{?}{=} 2$$

$$2 = 2 \quad \text{It checks.}$$

7. Solve. Check your solutions.

a. $\quad \frac{1}{3}x = -15$	**b.** $\quad -\frac{1}{6}x = -20$	**c.** $\quad -\frac{1}{4}x = 18$
d. $\quad -2 = -\frac{1}{9}x$	**e.** $\quad -21 = \frac{1}{8}x$	**f.** $\quad \frac{1}{12}x = -7 + 5$

Word Problems

Example 1. The area of a rectangle is 195 m². One side measures 13 m. How long is the other side?

To write the equation, you need to remember that the area of any rectangle is calculated as **area = side · side**.

That relationship gives us our equation. We simply substitute the known area and the length of the known side into the equation and represent the length of the unknown side by some variable:

$$195 = s \cdot 13$$

Then we rewrite the expression $s \cdot 13$ in the usual way, where the coefficient (13) comes first and the variable(s) last. We also flip the sides of the equation, so the unknown is on the left: $13s = 195$.

Solution:

$$13s = 195$$

$$\frac{13s}{13} = \frac{195}{13}$$ We divide both sides by 13 and simplify.

$$s = 15$$ This is the solution.

Check:

$$13 \cdot 15 \stackrel{?}{=} 195$$

$$195 = 195 \checkmark$$

For each given situation, write an equation and solve it. The problems themselves are simple, and you could solve them without writing an equation, but it is important to practice writing equations! You need to learn to write equations for simple situations now, so you will be able to write equations for more complex situations later on.

1. The perimeter of a square is 456 cm. How long is one side?

    Equation:

2. The area of a rectangular park is 4,588 square feet.
    One side measures 62 feet. How long is the other side?

    Equation:

3. John bought some boxes of screws for $15 each and paid a total of $165.
   How many boxes did he buy?

   Equation:

4. A candle burned at a steady rate of 2 cm per hour for 4 hours.
   Now it is only 6 cm long. How long was the candle at first?

   Equation:

5. A baby dolphin is only 1/12 as heavy as his mother. The baby weighs
   15 kilograms. How much does the mother dolphin weigh?

   Equation:

Example 2. Solve the equation $45 + y + 82 + 192 = 374$ using a bar model and also using an equation.

Bar model:

For the bar model, we draw the addends as parts of the total, which we indicate with a double-headed arrow.

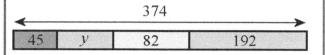

To solve it, we subtract all the known parts from 374:

$y = 374 - 45 - 82 - 192$

$y = 55$

Equation:

$$45 + y + 82 + 192 = 374$$
$$319 + y = 374$$
$$319 + y - 319 = 374 - 319$$
$$y = 55$$

First, add what you can on the left side.

Next, subtract 319 from both sides.

Check: $\quad 45 + 55 + 82 + 192 \stackrel{?}{=} 374$

$$374 = 374 \checkmark$$

6. Using both a bar model and an equation, solve the equation $21 + x + 193 = 432$.

Bar model:	Equation:

7. Using both a bar model and an equation, solve the equation $495 + 304 + w + 94 = 1{,}093$.

Bar model:	Equation:

Constant Speed

If an object travels with a constant speed, we have three quantities to consider: *speed or velocity* (*v*), *time* (*t*), and *distance* (*d*). The formula $d = vt$ tells us how they are interconnected.

Does that formula make sense?

Let's say John rides his bicycle at a constant speed of 12 km per hour for four hours. How far can he go? The formula says you multiply the speed (12 km/h) by the time (4 h) to get the distance (48 km). So the formula does make sense — that is how our "common sense" tells us to calculate it also.

Example 1. A boat travels at a constant speed of 15 km/h. How long will it take the boat to go a distance of 21 km?

The problem gives us the speed and the distance. The time (*t*) is unknown.

We can solve the unknown time by using the formula $d = vt$. We simply substitute the given values of *v* and *d* in it, and we will get an equation that we can solve:

$$d = v \, t$$
$$\downarrow \quad \downarrow$$
$$21 = 15 \, t$$

To make it easier, we will leave off the units while solving the equation. We can do that, since both the velocity and the distance involve kilometers.

Next, we simply solve this equation:

$21 = 15t$ Flip the sides.

$15t = 21$ Divide both sides by 15.

$\dfrac{15t}{15} = \dfrac{21}{15}$ The 15 in numerator and denominator cancel.

$t = 1 \, 6/15$

The final answer $t = 1 \, 6/15$ is in *hours*, because the unit for speed was kilometers per <u>hour</u>.

Let's work on that answer a bit more. First, it simplifies to 1 2/5 hours. Then let's change 2/5 of an hour to minutes.

How much is 1/5 of an hour? That's right, it is 12 minutes. And 2/5 of an hour is 24 minutes. So the final answer is 1 hour, 24 minutes.

You may use a calculator for all the problems in this lesson.

1. Use the formula $d = vt$ to solve the problems.

a. A caterpillar crawls along at a constant speed of 20 cm/min. How long will it take it to travel 34 cm?

$$d = v \quad t$$
$$\downarrow \quad \downarrow \quad \downarrow$$

b. Father leaves at 7:40 a.m. to drive 20 miles to work. If his average speed is 48 mph, when will he arrive at work?

$$d = v \quad t$$
$$\downarrow \quad \downarrow \quad \downarrow$$

How to change hours and minutes into fractional and decimal hours and vice versa	
Example 2. Change 14 minutes into hours. Since there are 60 minutes in an hour, 14 minutes is simply 14/60 of an hour. This simplifies to 7/30 hours. You can also change it into a decimal by dividing to get 0.233 hours (rounded to three decimals).	**Example 3.** Change 4.593 hours into hours and minutes. How many minutes are in the decimal part? Since one hour is 60 minutes, 0.593 hours is 0.593 · 60 minutes = 35.58 minutes ≈ 36 minutes. So 4.593 hours ≈ 4 hours 36 minutes.

2. Convert the given times into hours in decimal format. Round your answers to three decimal digits.

a. 35 minutes	**b.** 44 minutes
c. 2 h 16 min	**d.** 4 h 9 min

3. Give these times in hours and minutes.

a. 2.4 hours	**b.** 0.472 hours
c. 3 3/5 hours	**d.** 16/50 hours

4. The average speed of a bus is 64 km/hour. What distance can it travel in 4 hours and 15 minutes?

5. Sam is an athlete who can run 10 miles in an hour. How long will it take him to run home from the shopping center, a distance of 2.4 miles?

6. A train traveled a distance of 360 miles between two towns so that in the first half of the distance, its average speed was 90 mph, and in the second half, only 75 mph. How long did it take to travel from the one town to the other?

7. Elijah wants to use the extra time between classes for exercising. He plans to jog for 25 minutes in one direction, turn, and jog back to school. What is the distance Elijah can jog in 25 minutes if his average jogging speed is 6 mph?

Example 4. Andy drove from his home to his workplace, which was 22 miles away, in 26 minutes. What was his average speed?

The average speed of cars is usually given in miles per hour or kilometers per hour. This unit of speed actually gives us a **formula for calculating speed**:

$$\begin{array}{ccccc} \text{speed} & \text{is} & \text{miles} & \text{per} & \text{hour} \\ \downarrow & \downarrow & \downarrow & \downarrow & \downarrow \\ \text{velocity} & = & \text{distance} & / & \text{time} \end{array}$$

(The word "per" indicates division.)

In symbols, $v = \dfrac{d}{t}$.

His average speed is therefore

$$v = \frac{d}{t} = \frac{22 \text{ miles}}{26 \text{ minutes}} = \frac{11}{13} \text{ miles per minute}$$

$$\approx 0.846 \text{ miles per minute}.$$

The problem is that the average speed is usually given in miles per *hour*, not miles per minute. How can we fix that?

One way is to multiply our answer by 60. Doing that, we get 0.846 mi/min · 60 min/hr = 50.76 ≈ 51 mph.

Another way is to change the original time of 26 minutes into hours before using the formula.

Now 26 minutes is simply 26/60 hours, which simplifies to 13/30 hours. We can write

$$v = \frac{22 \text{ miles}}{13/30 \text{ hours}}$$

This is a **complex fraction**: a fraction that has another fraction in the numerator, denominator, or both.

One way to calculate its value with a calculator is to use parentheses and input it as 22 ÷ (13 ÷ 30). Check out what happens if you input it as 22 ÷ 13 ÷ 30.

Tip: Instead of parentheses, you can use the reciprocal button (1/x) on your calculator. First calculate the value of the fraction inverted (upside-down): $\dfrac{13/30}{22}$. This is the **reciprocal** of the fraction. You can input it as 13 ÷ 30 ÷ 22.

Once you've calculated the reciprocal, push the 1/x button to convert the reciprocal into the answer that you want.

8. Find the average speed in the given units.

   a. A duck flies 3 miles in 6 minutes.
   Give your answer in miles per hour.

   b. A lion runs 900 meters in 1 minute.
   Give your answer in kilometers per hour.

   c. Henry sleds 75 meters down the hill in 1.5 minutes.
   Give your answer in meters per second.

   d. Rachel swims 400 meters in 32 minutes.
   Give your answer in kilometers per hour.

9. Jake's grandparents live 150 km away from his home. One day it took him 2 h 14 min to get there and 1 h 55 min to come back home. In the questions below, round your answers to one decimal digit.

   a. What was his average speed going there?
   Hint: Change the time in hours and minutes into decimal hours.

   b. What was his average speed coming back?

   c. What was his overall average speed for the whole trip?

10. Another day Jake visited his grandparents again. Let's say that, because of the traffic, Jake achieved an average speed of 75 km/h going there but an average speed of only 65 km/h coming back. How much longer did Jake spend driving home from his grandparents' place than going there?

How to remember the formula $d = vt$	**Just solve for d from the formula $v = d/t$**
I will show you how to *derive* that formula! Then you do not really have to memorize it. But you *do* need to remember the formula $v = d/t$. You can remember *that* formula with the trick I explained earlier—by thinking of the common unit for measuring speed (miles per hour): speed is miles per hour ↓ ↓ ↓ ↓ ↓ velocity = distance / time In symbols: $v = d/t$.	$v = \dfrac{d}{t}$ We want d alone, so we multiply both sides by t. $vt = \dfrac{d}{t} \cdot t$ The t's on the right side cancel. $vt = d$ We have it! Turning it around we get $d = vt$, which is the most common formula to show how velocity (v), time (t), and distance (d) are related.

11. Compare the average speeds to find which bird is faster: a seagull that flies 10 miles in 24 minutes or an eagle that flies 14 miles in half an hour?

12. A train normally travels at a speed of 120 km per hour. One day, the conditions were so icy and cold that it had to slow down to travel safely. So the train traveled the first half of its 160-km journey at half its normal speed. Then the weather improved, and the train was able to go faster again. It sped the remaining distance at twice its normal speed to make up time.

   a. How long did the train take to travel the whole distance (160 km)?

   b. How long would the train have taken if it had traveled the whole trip at its normal speed?

   c. What was the train's average speed for the trip on this cold and icy day?

13. How long will it take Charlotte to ride her bike from the music store to her home—a distance of 4.5 km—if she rides 1/3 of it at 12 km/h and the rest at 15 km/h?

The next problems are more challenging.

14. Your normal walking speed is 6 miles per hour. One day you walk slowly, at 3 miles per hour, half the distance from home to the swimming pool. Can you now make up for your slow walking by walking the remaining distance at double your normal speed?

    (Hint: Make up a distance between your home and the swimming pool for an example calculation. Choose an easy number.)

15. An airplane normally flies at a speed of 1,000 km/h. Due to some turbulence, it has to travel at a lower speed of 800 km/h for the first 40 minutes of a 1600-km trip. How fast should it fly for the rest of the trip so as to make up for the lost time?

    (Hint: You will also need to calculate the normal traveling time for this trip.)

Chapter 3 Mixed Review

1. Write an expression.

   a. 10 less than x squared.

   b. The quotient of 154 and k cubed.

   c. The quantity x plus 2 to the fifth power.

   d. x plus 2 to the fifth power.

2. The sides of a square are $(x + 2)$ long.

   a. Sketch the square.

   b. Write an expression for the area of the square.

   c. Write an expression for the perimeter of the square.

   d. Evaluate your expression for the area of the square when $x = 1.5$.

3. Draw a number line jump for each addition or subtraction.

   a. $-2 + 6 =$ _____

   b. $-3 - 5 =$ _____

4. Draw counters for the addition $3 + (-5)$. Explain how to perform the addition using the counters.

5. Solve.

   | **a.** $89 + (-35) =$ | **b.** $-45 + (-29) =$ | **c.** $-78 + 60 =$ |

6. Change each addition into a subtraction or vice versa. Then solve whichever is easier. Sometimes changing the problem will not make solving it easier, but the aim of this exercise is to practice making the change.

   | **a.** $-2 + (-18)$ ↓ ___ − ___ = ___ | **b.** $56 - (-34)$ ↓ ___ + ___ = ___ | **c.** $-14 + (-24)$ ↓ ___ − ___ = ___ | **d.** $2 + 9$ ↓ ___ − ___ = ___ |

7. Write comparisons using > , < , and integers. Include the units, too.

   a. The temperature at the North Pole is −34 degrees Celsius, whereas in New York, it is −8 degrees Celsius.

   b. The total electric charge of 12 electrons is −12e.
   The total electric charge of 3 protons is +3e.

8. Name the property of arithmetic illustrated by the equation $2x = x \cdot 2$.

9. Evaluate the expression $|a - b|$ for the given values of a and b. Check that the answer you get is the same as if you had used a number line to figure out the distance between the two numbers.

a. a is 8 and b is 54	b. a is −12 and b is −5

10. Describe a situation where one person has a positive account balance and another has a negative balance, and the one person's balance is $30 more

11. Use the distributive property "backwards" to write the expression as a product.

   a. $42s + 28 = $ ____ (____ + ____)   b. $54z - 18 = $ ____ (____ − ____)

12. Solve the equations.

a. $\dfrac{x}{-5} = 35$ $x = $ _____	b. $\dfrac{35}{y} = -5$ $y = $ _____	c. $5z = -35$ $z = $ _____

13. Write the equation and then solve it using "guess and check." Each root is between −20 and 20.

a. 2 plus 14 equals x minus 1
b. x cubed equals 27

14. Add or subtract.

| a. (−9) + (−18) = _____ | b. −21 − (−3) = _____ | c. 17 − 51 = _____ |

15. Give a real-life situation for the sum 3 + (−10).

16. Simplify.

    a. $|-2|$ b. $-(-2)$ c. $-|2|$ d. -0

17. Find the value of the expressions when $x = -2$ and $y = 8$.

a. $5x^2$	b. $-5y + 6$	c. $-(y + x)$

18. Jeremy is 2 years older than Larry. Write an expression for Larry's age, if Jeremy is y years old.

19. Here is a growing pattern. Draw the steps 4 and 5 and answer the questions.

  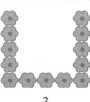

Step   1   2   3

a. How do you see this pattern grow?

b. How many flowers will be in step 39?

c. In step n?

Chapter 3 Review

1. Solve. Check your solutions.

a.	$x + 7 = -6$	**b.**	$-x = 5 - 9$
c.	$2 - x = -8$	**d.**	$2 - 6 = -z + 5$
e.	$\dfrac{x}{11} = -12$	**f.**	$\dfrac{q}{-3} = -40$
g.	$100 = \dfrac{c}{-10}$	**h.**	$\dfrac{a}{5} = -10 + (-11)$

2. *Write an equation for the problem. Then solve it.*

Alex bought three identical solar panels and paid a total of $837. How much did one cost?

Equation:

3. *Write an equation for the problem. Then solve it.*

   Andrew pays 1/7 of his salary in taxes. If he paid $187 in taxes, how much was his salary?

   Equation:

4. Use the formula $d = vt$ to solve the problem.

If you can bicycle at a speed of 20 km/h, how long will it take you to bicycle from the shopping center to a dentist's office, a distance of 1.2 km?	$d = v \quad t$ $\downarrow \quad \downarrow \quad \downarrow$

5. Taking a bus, Emily can get to the community center that is 1.5 km from her home in 3 minutes. What is the average speed of the bus, in kilometers per hour?

6. Ed skates on his skateboard to school, which is 2 miles away. He travels half of the distance at a speed of 12 mph and the rest at a speed of 15 mph. How long does it take him to get to school?

Chapter 4: Rational Numbers
Introduction

In this chapter we study *rational* numbers, which are numbers that can be written as a *ratio* of two integers. All fractions and whole numbers are rational numbers, and so are percents and decimals (except non-ending non-repeating decimals). Obviously, students already know a lot about rational numbers and how to calculate with them. Our focus in this chapter is to extend that knowledge to negative fractions and negative decimals.

The first lesson presents the definition of a rational number, how to convert rational numbers back and forth between their fractional and decimal forms, and a bit about repeating decimals (most fractions become repeating decimals when written as decimals). The next lesson deals with adding and subtracting rational numbers, with an emphasis on adding and subtracting negative fractions and decimals.

The next two lessons are about multiplying and dividing rational numbers. The first of the two focuses on basic multiplication and division with negative fractions and decimals. The second of the two compares multiplying and dividing in decimal notation to multiplying and dividing in fraction notation. Students come to realize that, though the calculations—and even the answers—may look very different, the answers are equal. The lesson also presents problems that mix decimals, fractions, and percents, and deals with real-life contexts for the problems and the importance of pre-estimating what a reasonable answer would be.

The lesson *Many Operations with Rational Numbers* reviews the order of operations and applies it to fraction and decimal problems with more than one operation. It also presents a simple method to solve **complex fractions** (fractions that contain another fraction, either in the numerator, in the denominator, or in both).

After a lesson on scientific notation, the instructional portion of the chapter concludes with two lessons on solving simple equations that involve fractions and decimals.

You can find matching videos for some of the topics in this chapter at
https://www.mathmammoth.com/videos/prealgebra/pre-algebra-videos.php#rational

The Lessons in Chapter 4

	page	span
Rational Numbers	116	*8 pages*
Adding and Subtracting Rational Numbers	124	*6 pages*
Multiply and Divide Rational Numbers 1	130	*4 pages*
Multiply and Divide Rational Numbers 2	134	*7 pages*
Many Operations with Rational Numbers	141	*4 pages*
Scientific Notation	145	*3 pages*
Equations with Fractions	148	*5 pages*
Equations with Decimals	153	*3 pages*
Chapter 4 Mixed Review	156	*2 pages*
Chapter 4 Review	158	*5 pages*

Helpful Resources on the Internet

You can also access this list of links at https://links.mathmammoth.com/gr7ch4

Rational Numbers - Video Lessons by Maria
A set of free videos that teach the topics in this book - by the author.
https://www.mathmammoth.com/videos/prealgebra/pre-algebra-videos.php#rational

RATIONAL NUMBERS

Compare Positive and Negative Decimals
Choose the correct inequality or equals sign from the dropdown box between each set of two numbers.
https://www.transum.org/software/SW/Starter_of_the_day/Students/Inequalities.asp?Level=2

Compare Rational Numbers
Practice comparing decimals, percents, fractions, and mixed numbers in this interactive exercise.
https://bit.ly/Compare-Rational-Numbers

Writing Fractions as Repeating Decimals
Answer multiple-choice questions about fractions and repeating decimals.
https://bit.ly/Writing-Fractions-as-Repeating-Decimals

Terminating vs. Repeating Decimals Game
A card game that practices repeating and terminating decimals. Several fun twists to score extra points! This game costs $1 (per download).
https://www.teacherspayteachers.com/Product/Terminating-VS-Repeating-Decimals-Game-425199

Converting Fractions to Terminating and Repeating Decimals Worksheets
Use these printable worksheets as tests, practice assignments, or teaching tools.
https://www.math-drills.com/fractions/fractions_convert_to_decimal_001.php

Converting Repeating Decimals to Fractions
A lesson that explains the method for writing repeating decimals as fractions.
https://www.basic-mathematics.com/converting-repeating-decimals-to-fractions.html

Classifying Numbers
Drag the given numbers to the correct sets. This chapter of Math Mammoth does not teach about square roots and irrational numbers, but you can probably do these activities, if you note that most square roots are irrational, and that the set of whole numbers is {0, 1, 2, 3, 4, ...}.
https://www.softschools.com/math/classifying_numbers/

Rational Numbers on a Number Line
Place the given rational numbers to the correct places on the number line.
https://www.geogebra.org/m/ZUk32S3x

ADD AND SUBTRACT RATIONAL NUMBERS

Adding and Subtracting Rational Numbers Worksheets
Generate a worksheet for adding and subtracting negative fractions and decimals.
https://www.math-aids.com/Algebra/Algebra_1/Basics/Add_Sub_Rational.html

Add Decimals Quiz
Reinforce your decimal addition skills with this 10-question online quiz.
https://www.thatquiz.org/tq-3/?-j1i1-lk-p0

Add Fractions Quiz
Add the fractions, and express the answer as a simple fraction in lowest terms in this 10-question quiz.
https://www.thatquiz.org/tq-3/?-j1gh-la-p0

Adding and Subtracting Rational Numbers Quiz
Practice addition and subtraction of rational numbers with this interactive quiz.
https://www.softschools.com/quizzes/math/adding_and_subtracting_rational_numbers/quiz3284.html

Adding and Subtracting Negative Fractions
Practice adding and subtracting positive and negative fractions with this interactive online activity.
https://bit.ly/Adding-and-Subtracting-Negative-Fractions

Adding and Subtracting Rational Numbers
Practice adding and subtracting negative fractions, decimals, and percents with this interactive online exercise.
https://bit.ly/Adding-and-Subtracting-Rational-Numbers

Equations with Fractions
Practice solving linear equations that contain fractions in this multi-level interactive exercise.
https://www.transum.org/Maths/Exercise/Algebra/Fraquations/

MULTIPLY AND DIVIDE RATIONAL NUMBERS

Multiplying Positive and Negative Fractions
Practice multiplication of positive and negative fractions with this short interactive quiz.
https://bit.ly/Multiplying-Positive-and-Negative-Fractions

Multiply and Divide Rational Numbers Quiz
A multiple-choice quiz of five questions.
https://www.softschools.com/quizzes/math/multiply_rational_numbers/quiz3285.html

Multiply and Divide Fractions
Practice multiplying and dividing fractions with positive numbers with this interactive exercise.
http://www.onemathematicalcat.org/algebra_book/online_problems/md_fractions.htm#exercises

Divide Positive and Negative Fractions
Practice dividing fractions. The fractions in these problems may be positive or negative.
https://bit.ly/Divide-Positive-and-Negative-Fractions

Hexingo — Operations with Rational Numbers
Practice adding, subtracting, and multiplying a mixture of fractions and decimals (some are negative) in this fun game!
https://www.mathmammoth.com/practice/rational-numbers

Fractions & Decimals Matching Game
Practice converting fractions to decimals while also uncovering a hidden picture in this fun matching game!
https://www.mathmammoth.com/practice/fractions-decimals

Multiplying and Dividing Negative Numbers Word Problems—from Khan Academy
Practice matching situations to multiplication and division expressions and equations.
https://bit.ly/multiply-divide-negative-numbers-problems

Simplify Complex Fractions
Practice simplifying complex fractions with this interactive online activity from Khan Academy.
https://bit.ly/Simplify-Complex-Fractions

Simplify Complex Fractions Practice (without mixed numbers)
Practice simplifying complex fractions in this interactive online exercise.
https://www.mathmammoth.com/practice/complex-fractions#mixed=0&questions=10

Simplify Complex Fractions Practice (with mixed numbers)
Practice simplifying complex fractions with this interactive online quiz.
https://www.mathmammoth.com/practice/complex-fractions#mixed=1&questions=10

SCIENTIFIC NOTATION

Scientific Notation Quiz
Write numbers in scientific notation, and vice versa. You can also modify the quiz parameters.
https://www.thatquiz.org/tq-c/?-j820-l6-p0

Scientific Notation
Practice converting numbers written in standard form to scientific notation and vice versa in this interactive online quiz.
https://bit.ly/Scientific-Notation

EQUATIONS / GENERAL

One-step Multiplication & Division Equations: Fractions & Decimals
Reinforce your equation solving skills with this short interactive online activity.
https://bit.ly/One-Step-Multiplication

Fraction Four Game
Choose "algebra" as the question type to solve equations that involve fractions in this connect-the-four game.
http://www.shodor.org/interactivate/activities/FractionFour/

One-Step Equations with Fractions
This algebra 1 worksheet will produce one step problems containing fractions.
https://www.math-aids.com/Algebra/Algebra_1/Equations/One_Step_Fractions.html

One-Step Equations: Fractions and Decimals
Practice solving equations in one step by multiplying or dividing a number from both sides.
https://bit.ly/One-Step-Equations

7th Grade Numbers and Operations Jeopardy Game
The questions in this game range from absolute value to different operations with rational numbers.
https://www.math-play.com/Integers-Jeopardy/integers-jeopardy-fun-game_html5.html

Rational Numbers

If you can write a number as a *ratio of two integers*, it is a **rational number**.   For example, 4.3 is a rational number because we can write it as the ratio $\frac{43}{10}$ or 43:10.   Note: To represent rational numbers, we usually indicate the ratio with a fraction line rather than a colon.

Examples of rational numbers	Since −10 can be written as $\frac{-10}{1}$, it is a rational number. It can also be written as $\frac{10}{-1}$. Since 0.1 can be written as $\frac{1}{10}$, it is also a rational number. Since 3.24 can be written as $\frac{324}{100}$, it, too, is a rational number.

Negative fractions   The ratio of the integers 7 and −10 gives us the fraction $\frac{7}{-10}$. As we studied earlier, we usually write this as $-\frac{7}{10}$ and read it as "negative seven tenths."

Obviously, all fractions, whether negative or positive, are rational numbers.

Negative fractions give us negative decimals.   For example, $-\frac{8}{10}$ is written as a decimal as −0.8, and $-5\frac{21}{100} = -5.21$.

You can write a rational number as a ratio of two integers in many ways.   For example, the decimal −1.4 can be written as a ratio of two integers in all these ways (and more!): $-1.4 = \frac{-14}{10} = \frac{-28}{20} = \frac{28}{-20} = \frac{42}{-30} = \frac{-42}{30} = \frac{-7}{5}$ So −1.4 is *definitely* a rational number! ☺ But the same holds true for all rational numbers—you can always write them as a ratio of two integers in multitudes of ways.

1. Write these numbers as a ratio (fraction) of two integers.

a. 6	**b.** −100	**c.** 0	**d.** 0.21
e. −1.9	**f.** −5.4	**g.** −0.56	**h.** 0.022

2. Are all percents, such as 34% or 5%, rational numbers? Justify your answer.

3. Form a fraction from the two given integers. Then convert it into a decimal.

a. 8 and 5	b. −4 and 10	c. 89 and −100
d. −5 and 2	e. 91 and −1000	f. −1 and −4

4. Mark the fractions on the number line below: $-\dfrac{1}{2}$, $-\dfrac{7}{8}$, $-1\dfrac{5}{8}$, $-2\dfrac{1}{4}$, $-2\dfrac{3}{4}$

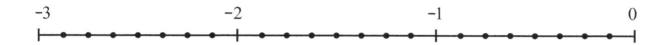

5. Write the fractions and mixed numbers marked by the arrows.

6. Mark the decimals on the number line: −0.11, −0.58, −0.72, −0.04

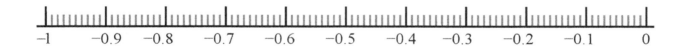

7. Sketch a number line from −3 to 0. Place tick marks at every tenth. Then mark the following numbers on your number line: −0.2, −1.5, −2.8.

Review: How to write decimals as fractions	• Simply copy all the digits from the decimal number that will form <u>the numerator</u> (omit the decimal point). • <u>The denominator</u> is always the power of ten with as many zeros as there are decimal digits in the number. **Example 1.** $3.0928 = \dfrac{30{,}928}{10{,}000}$ (Four decimal digits means the denominator is 10^4 or 10,000.)

You can also write decimals as mixed numbers, in which case you copy only the digits from the *decimal part* to be the numerator, and the whole-number part of the decimal becomes the whole-number part of the mixed number.

Example 2. $5.00447301 = 5\dfrac{447301}{100{,}000{,}000}$

Note: This method does *not* apply to non-ending decimals (such as 1.333333…)!

8. Write these decimals as fractions.

a. 0.3928	b. 1.028384	c. 0.0043928
d. −0.00584	e. −9.2031	f. 1018.2939

9. Write these decimals as mixed numbers.

a. 2.0038	b. 7.483901	c. 101.4832
d. −12.039	e. −4830.22	f. −8.028567

10. Let's also try this the other way around! Write the fractions as decimals. Reminder: The number of zeros in the denominator tells you the number of decimal digits you need.

a. $-\dfrac{8}{10{,}000}$	b. $\dfrac{3{,}107}{100}$	c. $8\dfrac{938}{100{,}000}$
d. $-\dfrac{553{,}911}{10{,}000}$	e. $\dfrac{3{,}912{,}593}{1000}$	f. $\dfrac{45{,}101}{1{,}000{,}000}$

11. Write these rational numbers as ratios of two integers (fractions) in a lot of different ways.

a. $-2 = -\dfrac{2}{1} =$

b. $0.6 = \dfrac{6}{10} =$

Writing fractions as decimals

You have just seen that it is easy to rewrite a fraction as a decimal when the denominator is a power of ten. However, when it is not (which is most of the time), simply treat the fraction as a division and divide. You will get either a **terminating decimal** or a non-terminating **repeating decimal**. See the examples below.

Example 3. Write $\frac{31}{40}$ as a decimal.

This division **terminates** (comes out even) after just three decimal digits.

We get $\frac{31}{40} = 0.775$. This is a **terminating decimal**.

```
     0 0.7 7 5
40) 3 1.0 0 0 0
    -2 8 0
      3 0 0
    -  2 8 0
        2 0 0
      -2 0 0
            0
```

Example 4. Write $\frac{18}{11}$ as a decimal.

We write 18 as 18.0000 in the long division "corner" and divide by 11. Notice how the digits "63" in the quotient and the remainders 40 and 70 start repeating.

So $\frac{18}{11} = 1.636363...$

We can use an ellipsis (three dots, or "…") to indicate that the decimal is non-terminating. A better notation is to draw a **bar** (a line) over the digits that repeat: $1.636363... = 1.\overline{63}$.

This number is called a **repeating decimal** because the digits "63" repeat forever!

```
     0 1.6 3 6 3
11) 1 8.0 0 0 0
   -1 1
      7 0
    -6 6
        4 0
      -3 3
          7 0
        -6 6
            4 0
          -3 3
              7
```

The decimal form of ANY rational number is either a terminating decimal or a repeating decimal.

Example 5. The repeating decimal 1.90510505050505… is written as $1.9051\overline{05}$. Notice that the bar marks only the digits that repeat ("05"). The digits "9051" aren't included.

Example 6. The decimal $0.05\overline{61}$ means the same as 0.056161616161...

Since it is a repeating decimal, it is a rational number. That means we can write it as a fraction. Which fraction?

We will not study this method in this grade, but just as a preview, here is how it is done:

In our decimal (let's call it *x* for short) we need to make the repeating digits go away. Here two digits repeat, so if we multiply *x* by 100, the digits repeat in 100*x* in the same place values that they do in *x*. (Why?) That means we can subtract 100*x* − *x* = 99*x* to make the repeating digits go away. Then if we divide both sides by 99, our decimal has become a fraction in ninety-ninths. All we have left to do is to find a number to multiply the numerator and denominator to make them both integers (preferably in the lowest terms). Like this:

$100x = 5.616161 ...$
$- x = 0.056161 ...$
_____
$99x = 5.56$
$x = 5.56/99 = 556/9900$
$ = 139/2475$

You can divide these fractions on a calculator to check them. Other repeating decimals can be figured in a similar way. For example, if three digits repeat in *x*, you will need to calculate 1000*x* − *x* = 999*x*. And so on.

You can find more about the method for writing repeating decimals as fractions on web pages like:
https://www.basic-mathematics.com/converting-repeating-decimals-to-fractions.html

12. Write each decimal using a line over the repeating part.

| a. 0.09090909... | b. 5.6843434343... | c. 0.19866666666... |

13. Do it the other way around: write the repeating digits several times followed by an ellipsis (three dots).

   a. $0.0\overline{887}$ b. $0.2\overline{456}$ c. $2.1\overline{7234}$

14. a. Is $0.98989898... = 0.\overline{98}$ a rational number?

   b. Is $2.064\overline{241}$ a rational number?

   c. Are all repeating decimals rational numbers?

> **Example 7.** The decimal 0.095 is a terminating decimal, but we *could* write it with an unending decimal expansion if we write zeros for all the decimal places after thousandths:
>
> $$0.095 = 0.095000000000...$$
>
> In other words, we can think of it as repeating the digit zero. In that sense, $0.095 = 0.095\overline{0}$. However, as you know, we normally write terminating decimals without the extra zeros.

15. Which decimal is greater?

a. Which is more, $0.\overline{3}$ or 0.3? How much more?	b. Which is more, $0.\overline{55}$ or $0.\overline{5}$? How much more?
c. Which is more, $0.45\overline{0}$ or 0.45? How much more?	d. Which is more, $0.\overline{12}$ or 0.12? How much more?

16. Write as decimals, using a line over the repeating part (if any). Use long division.

| a. $\frac{2}{3}$ | b. $1\frac{1}{3}$ | c. $\frac{8}{9}$ |

16. Write as decimals, using a line over the repeating part (if any). Use long division.

d. $2\frac{7}{16}$	e. $\frac{19}{24}$	f. $\frac{1}{6}$

Example 8. Write the decimal expansion of $\frac{8}{31}$ to five decimal digits.

We cannot see any repeating pattern in the long division in the first six decimals. Therefore, we stop the division after six decimals, and round the number to five decimals.

We get $\frac{8}{31} \approx 0.25806$.

Now, there *is* a pattern in the decimal digits. (How do we know that? Because 8/31 is a rational number.) But the pattern is 15 digits long! You can find it with a calculator that shows more than 15 digits, such as one on a computer.

(The calculator gives the decimal expansion as 0.25806451612903225806451612903226. Can you find the repeating part?)

```
         0.2 5 8 0 6 4
    31 ) 8.0 0 0 0 0 0
        -6 2
         1 8 0
        -1 5 5
           2 5 0
          -2 4 8
             2 0
            -  0
             2 0 0
            -1 8 6
               1 4 0
              -1 2 4
                 1 6
```

17. **a.** Find the repeating pattern in the decimal expansion of 2/17 using a calculator.

   Hint: if your calculator doesn't show enough decimals, try these online calculators:

   https://keisan.casio.com/calculator

   https://www.mathsisfun.com/calculator-precision.html

   b. Find the repeating pattern in the decimal expansion of 5 17/21 using a calculator.

18. Write in decimal form. Use long division, and calculate each answer to at least six decimal places. If you find a repeating pattern, give the repeating part. If you don't, round your answer to five decimals.

a. $2\frac{5}{18}$	b. $0.23 \div 4$	c. $0.76 \div 11$

Are there any numbers that are *not* rational? Are there unending decimals that *don't* repeat in a pattern?

Yes, there are. For example, if you divide the circumference of an ideal circle by its diameter, you get a number that we denote as Pi. Pi *cannot* be written as a ratio of two integers. It is an **irrational number**. Its decimal expansion goes on forever *without* any repeating pattern. Here are the first few digits of it:

Pi = 3.1415926535897932384626433832795028841971693993751058209…

Another example is the square root of 2 (the number whose square is 2). It is close to 1.41421356, but once again, its decimal expansion goes on forever without any repeating pattern. It can be proven that it cannot be written as a fraction, so it is irrational.

Mathematicians have found many other irrational numbers as well. In fact, there are *more* irrational numbers than there are rational ones. However, our normal daily life revolves around rational numbers.

19. Patterns! Write as decimals. Use a notebook for long divisions. Not all of these are repeating decimals, but those that are have a pattern in their repeating parts! The first one is done for you.

a.	b.	c.	d.
$\frac{1}{3} = 0.\overline{3}$	$\frac{1}{9} =$	$\frac{1}{4} =$	$\frac{1}{6} =$
$\frac{2}{3} = 0.\overline{6}$	$\frac{2}{9} =$	$\frac{2}{4} =$	$\frac{2}{6} =$
$\frac{3}{3} = 1$	$\frac{3}{9} =$	$\frac{3}{4} =$	$\frac{3}{6} =$
$\frac{4}{3} = 1.\overline{3}$	$\frac{4}{9} =$	$\frac{4}{4} =$	$\frac{4}{6} =$
$\frac{5}{3} = 1.\overline{6}$	$\frac{5}{9} =$	$\frac{5}{4} =$	$\frac{5}{6} =$
$\frac{6}{3} = 2$	$\frac{6}{9} =$	$\frac{6}{4} =$	$\frac{6}{6} =$

e.	f.	g.	h.
$\frac{1}{7} = 0.\overline{142857}$	$\frac{1}{8} =$	$\frac{1}{5} =$	$\frac{1}{11} =$
$\frac{2}{7} =$	$\frac{2}{8} =$	$\frac{2}{5} =$	$\frac{2}{11} =$
$\frac{3}{7} =$	$\frac{3}{8} =$	$\frac{3}{5} =$	$\frac{3}{11} =$
$\frac{4}{7} =$	$\frac{4}{8} =$	$\frac{4}{5} =$	$\frac{4}{11} =$
$\frac{5}{7} =$	$\frac{5}{8} =$	$\frac{5}{5} =$	$\frac{5}{11} =$
$\frac{6}{7} =$	$\frac{6}{8} =$	$\frac{6}{5} =$	$\frac{6}{11} =$
$\frac{7}{7} =$	$\frac{7}{8} =$	$\frac{7}{5} =$	$\frac{7}{11} =$

When the fraction 1/3 is written as a decimal, it is 0.33333…. This could be rounded to three decimals (0.333), or to six decimals (0.333333), or to any other number of decimals.

Find the *difference* between $0.\overline{3}$ rounded to five decimals and $0.\overline{3}$ rounded to only two decimals.

Adding and Subtracting Rational Numbers

In this lesson you will add and subtract rational numbers (both fractions and decimals), with emphasis on adding and subtracting *negative* rational numbers. It is fairly easy because **adding and subtracting negative rational numbers works the same way as adding and subtracting integers**.

Adding several negative numbers

How do you add if all the addends are negative? For example, to add −10 + (−5) + (−7) + (−9), we add the *absolute values* of the numbers, then give the answer as a negative. Lots of negatives add up to a large negative!

Since 10 + 5 + 7 + 9 = 31, the answer to the original sum is −31.

It works the same way with negative rational numbers. For example, to find the sum −0.2 + (−1.9) + (−2.05) + (−0.78), we first add the absolute values, which gives us 4.93. The answer to the original sum is −4.93.

```
  0.2
  1.9
  2.05
+ 0.78
------
  4.93
```

Adding one positive and one negative number

Do you remember how to add −14 + 11 or 25 + (−9)?

When you have one negative and one positive number, check which side "wins," so to speak. In other words, check whether the positive or the negative number has the greater absolute value.

The next step is to actually subtract the absolute values of the two numbers to determine, so to speak, by how much the winning side wins. To do that you need to find the difference in their absolute values. It is kind of funny to think that to find the answer to an addition problem you need to subtract, but it is so!

Example 1. Add 19.58 + (−42.3).

Compare the absolute values. Since 19.58 < 42.3, the negative side will "win" and our final answer will be negative. To find out by how much the negative side wins, we subtract the absolute values of the two numbers: 19.58 and 42.3. (See the subtraction on the right.)

Therefore, 19.58 + (−42.3) = −22.72.

```
   11 12
 3 ⧸1 ⧸2 10
   4 2.3 0
 − 1 9.5 8
 ---------
   2 2.7 2
```

1. Add.

| a. −0.9 + (−0.6) | b. −1.8 + 0.5 | c. −0.4 + 0.7 |

2. Add.

| a. −0.9 + (−0.67) + (−2.405) | b. −10.08 + 4.5 | c. −3.4 + 7.98 |

Adding fractions

To add fractions, you can follow the same procedure as for decimals, but there is also an easier way.

Example 2. Add $\frac{3}{4} + \left(-\frac{6}{7}\right)$. We could figure out which fraction has a larger absolute value, then subtract the smaller absolute value from the larger one, and so on. But the easier way is this: Simply add the fractions normally and treat the negative fraction $-\frac{6}{7}$ as $\frac{-6}{7}$. You will end up with an *integer addition* in the numerator.

Example. $\frac{3}{4} + \left(-\frac{6}{7}\right)$

$$\frac{3}{4} + \frac{-6}{7}$$

$$\frac{21}{28} + \frac{-24}{28}$$

$$\frac{21 + (-24)}{28} = \frac{-3}{28} = -\frac{3}{28}$$

3. Add the fractions.

a. $\frac{3}{5} + \left(-\frac{2}{3}\right)$

b. $\frac{1}{9} + \left(-\frac{6}{9}\right)$

c. $-\frac{3}{4} + \frac{2}{9}$

d. $-\frac{1}{6} + \frac{3}{8}$

e. $\frac{1}{8} + \left(-\frac{1}{3}\right)$

f. $\frac{9}{10} + \left(-\frac{2}{3}\right)$

Adding a mix of positive and negative numbers

1. First add all the positive numbers and all the negative numbers separately.

2. Lastly add these two totals as we did in the box, "Adding one positive and one negative number."

Example 3. Add $-9.5 + 2.4 + 0.5 + (-4.3) + (-0.8)$.

Adding all the negative decimals, we get $-9.5 + (-4.3) + (-0.8) = \mathbf{-14.6}$.

The positives total to $2.4 + 0.5 = \mathbf{2.9}$.

Lastly, we add those two totals: $\mathbf{2.9} + (\mathbf{-14.6}) = -11.7$.

4. Add.

a. $-0.5 + 0.6 + (-1.2) + (-1.4) + 1.6$	**b.** $-\$1.08 + (-\$4.30) + \$0.56 + \$0.99 + (-\$0.25)$
c. $\dfrac{1}{2} + \left(-\dfrac{1}{3}\right) + \left(-\dfrac{4}{3}\right) + \dfrac{1}{6}$	**d.** $\dfrac{5}{8} + \left(-\dfrac{1}{2}\right) + \dfrac{3}{4} + \left(-\dfrac{3}{2}\right) + \left(-\dfrac{7}{8}\right)$

Subtracting rational numbers
1. Sometimes you can use mental math and visualize a number line jump. **Example 4.** 1.2 − 1.5 Think of starting at 1.2 and moving towards the negatives. You will pass zero and go three tenths farther into the negative "zone." The answer is −0.3.
2. At other times, you may need to use the definition of subtraction: The difference of the numbers *a* and *b* is the <u>sum</u> of *a* and the opposite of *b*: $a - b = a + (-b)$ In other words, instead of subtracting a number, you add its opposite. **Example 5.** Solve $-\frac{11}{12} - \left(-\frac{1}{3}\right)$. We simply change the subtraction into addition to get $-\frac{11}{12} + \frac{1}{3}$. Then, $-\frac{11}{12} + \frac{1}{3} = -\frac{11}{12} + \frac{4}{12} = \frac{-11 + 4}{12} = \frac{-7}{12} = -\frac{7}{12}$.

5. Draw a number line jump for each addition or subtraction, just like you did with integers.

a. −1.4 − 0.8   **b.** 1 − 2.8

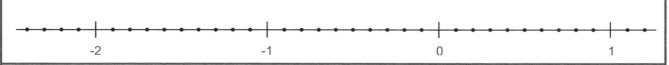

c. 1.1 − 2.2   **d.** −1.7 + 0.8

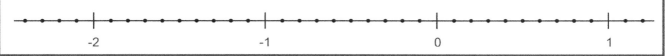

e. −0.9 + (−1.2)   **f.** −1.2 − (−0.5)

6. Find the distance between the two numbers. The number lines above can help.

   a. −0.8 and −2.2   **b.** 0.9 and −1.3   **c.** −1.1 and −0.4

7. Solve.

a. −1.2 − 0.6	**b.** 0.3 − 1.2	**c.** −0.8 − (−0.8)

8. Solve.

a. $0.97 - 1.67$	b. $-5.61 - 0.9$	c. $2.5 - (-4.2) + (-0.3)$

9. Solve.

a. $-\dfrac{2}{9} - \left(-\dfrac{2}{3}\right) - \dfrac{5}{9}$	b. $\dfrac{5}{8} + \left(-\dfrac{1}{8}\right) + \dfrac{1}{4} + \left(-\dfrac{1}{2}\right) + \left(-\dfrac{9}{8}\right)$

10. Mark has $5.50 in his piggy bank. He wants to buy an activity book that costs $7.90. His mom said he can owe her the part that he cannot pay now. Write a number sentence to represent Mark's money situation (balance) after the purchase.

11. Explain a real-life situation for the sum $-\$50 + (-\$12.90)$.

12. A weather station measures the temperature every hour. At 6 AM it was $-5.6°C$ and at 12 PM it was $-0.9°C$. How much warmer was it at noon than at 6 AM?

13. Do you remember how to find the distance between two numbers *a* and *b*? It is given by the expression |*a* − *b*|. Evaluate the expression |*a* − *b*| for the given values of *a* and *b*. Check that the answer you get is the same as if you had used a number line to figure out the distance between the two numbers.

a. *a* = 0.7 and *b* = −0.7 | − | =	**b.** *a* = −7.8 and *b* = −5.4
c. *a* = 1/4 and *b* = −3 1/4	**d.** *a* = −4/10 and *b* = −9/10

14. Which expressions can be used to find the distance between *x* and 2/3?

a. |*x* − 2/3|	**b.** *x* − 2/3	**c.** |*x* + 2/3|	**d.** |2/3 − *x*|	**e.** *x* − (−2/3)	**f.** |*x* − (−2/3)|

15. The table lists the average high and low temperatures for each month in the state of Alaska. It is based on the averages of daily high and low temperatures in 34 towns in Alaska.

	JAN	FEB	MAR	APR	MAY	JUN	JUL	AUG	SEP	OCT	NOV	DEC
Average High Temperature (°C)	−4.3	−8.9	−6.3	−8.5	4.4	10.5	16.1	17.4	16.2	11.4	3.1	−4.3
Average Low Temperature (°C)	−4.7	−16.5	−15.1	−14.7	−5.7	0.4	5.7	7.8	6.8	2.7	−4.1	−11.4
Difference												

Calculate the difference between the average high and low temperature for each month. Describe your findings.

In which month(s) is the temperature variation the greatest?

In which month(s) is the temperature variation the least?

16. Quarks are particles that are even smaller than protons and neutrons. In fact, protons and neutrons are thought to be made up of quarks. The two most common kinds of quarks (whimsically called "flavors") are "up" (**u**) quarks, with a charge of +2/3*e*, and "down" (**d**) quarks, with a charge of −1/3*e*, where *e* is the so-called "elementary charge" (defined as the charge on a proton).

   a. A **neutron** (n) is thought to be composed of one up and two down quarks (**u**+**d**+**d**). Based on that, what should the charge on a neutron be?

   b. A **proton** (p) is thought to be composed of two up quarks and one down quark (**u**+**u**+**d**). If that is so, then what should the charge on a proton be?

   c. A **pion** (pi meson, π+) is thought to be composed of an up quark and a down anti-quark (the opposite of a down quark) (**u**+(¯**d**)). Given that, what charge should a pion have?

   d. An **anti-pion** (pi anti-meson, π−) is thought to be composed of an up anti-quark (the opposite of an up quark) and a down quark ((¯**u**)+**d**). Assuming that is correct, what should the charge be on an anti-pion?

Multiply and Divide Rational Numbers 1

In real life we often combine **fractions, decimals, ratios, and percents** — rational numbers in different forms — in the same situation. You need to be able to easily calculate with them in their different forms.

In this lesson, we will concentrate on multiplying and dividing *decimals* and *fractions* because percentages are usually rewritten as decimals and ratios as fractions before calculating with them.

To multiply decimals

Shortcut: First multiply as if there were no decimal points. Then put the decimal point in the answer so that the number of decimal digits in the answer is the SUM of the number of the decimal digits in all the factors.

Example 1. Solve $-0.2 \cdot 0.09$.

Multiply $2 \cdot 9 = 18$. The answer will have three decimals *and* be negative (Why?), so the answer is -0.018.

Multiply fractions & mixed numbers	**Example 2.** $-\dfrac{4}{5} \cdot \left(-5\dfrac{1}{8}\right)$	
1. Change any mixed numbers to fractions.	$= -\dfrac{4}{5} \cdot \left(-\dfrac{41}{8}\right)$	A negative times a negative makes a positive, so we can drop the minus signs in the next step.
2. Multiply using the shortcut (multiply the numerators; multiply the denominators).	$= \dfrac{4 \cdot 41}{5 \cdot 8} = \dfrac{1 \cdot 41}{5 \cdot 2} = \dfrac{41}{10} = 4\dfrac{1}{10}$	

1. Write the rational numbers in their four forms.

	ratio		fraction		decimal		percent			ratio		fraction		decimal		percent
a.	2:5	=	$\dfrac{2}{5}$	=	0.4	=	40%	**d.**		=	$\dfrac{7}{20}$	=		=	
b.	3:4	=		=		=		**e.**		=		=		=	55%
c.	4:25	=		=		=		**f.**		=		=	0.85	=	

2. Multiply these in your head.

a. $0.1 \cdot 6.5$	**b.** $-0.08 \cdot 0.006$	**c.** $-0.09 \cdot 0.02$
d. $-0.2 \cdot (-1.6)$	**e.** $-0.8 \cdot 1.1 \cdot (-0.02)$	**f.** 0.8^2
g. $(-0.5)^2$	**h.** $(-0.2)^3$	**i.** $(-0.1)^5$

3. Multiply

a. $-\dfrac{1}{7} \cdot \left(-\dfrac{3}{8}\right)$	b. $\dfrac{1}{5} \cdot \left(-2\dfrac{1}{2}\right)$	c. $-\dfrac{2}{9} \cdot \dfrac{5}{6} \cdot \dfrac{3}{10}$
d. $-3\dfrac{1}{4} \cdot \dfrac{5}{2}$	e. $\dfrac{7}{18} \cdot \left(-\dfrac{12}{27}\right)$	f. $\dfrac{8}{7} \cdot \left(-\dfrac{3}{10}\right) \cdot 1\dfrac{1}{2}$

4. Multiply using the regular multiplication algorithm (write one number under the other).

a. $12.5 \cdot 2.5$	b. $-0.088 \cdot 0.16$
c. $-9.08 \cdot (-0.006)$	d. $24 \cdot (-0.0087)$

To divide decimals

1. If the divisor has no decimal digits, you can divide using long division "as is."

2. If the divisor does have decimal digits, multiply *both* the dividend and the divisor by the same number (usually a power of ten) to make the divisor into a whole number. Now with that whole number divisor, performing the long division has become straightforward.

Example 3. Solve $6 \div 0.5$ without a calculator.

Since 0.5 fits into 6 exactly twelve times, the answer is 12. So mental math was sufficient in this case.

Example 4. Solve $-92.91 \div 0.004$ without a calculator.

It may be easier to write the problem using a fraction line:

$$\frac{-92.91}{0.004} = \frac{-929.1}{0.04} = \frac{-9291}{0.4} = \frac{-92910}{4}$$

Notice how we multiply both the dividend and the divisor repeatedly by 10 until the divisor becomes a whole number (4). (You could, of course, simply multiply them both by 1,000 to start with.) Then we use long division.

The long division gives us the absolute value of the final answer, but we still need to apply the correct sign. So $-92.91 \div 0.004 = -23,227.5$.

Does this make sense? Yes. The answer has a very large absolute value because 0.004 is a very tiny number, and thus it "fits" into 92.91 multitudes of times.

```
        2 3 2 2 7.5
    4)9 2 9 1 0.0
      -8
       1 2
      -1 2
         0 9
         - 8
           1 1
         -  8
            3 0
           -2 8
              2 0
             -2 0
                0
```

5. Divide using mental math.

a. $-0.88 \div 4$	b. $8.1 \div 9$	c. $72 \div 10000$
d. $-1.6 \div (-0.2)$	e. $8 \div 0.1$	f. $0.8 \div (-0.04)$

6. Multiply both the dividend and the divisor by the same number so that you get a divisor that is a *whole number*. Then divide using long division. If necessary, round your answer to three decimal digits.

a. $27.6 \div 0.3$	b. $2.088 \div 0.06$

To divide fractions and mixed numbers	**Example 5.** $\dfrac{4}{5} \div \left(-2\dfrac{1}{2}\right)$
1. Change any mixed numbers to fractions.	$= \dfrac{4}{5} \div \left(-\dfrac{5}{2}\right)$
2. Divide using the shortcut. (Change the division into a multiplication by the reciprocal of the divisor.)	$= \dfrac{4}{5} \cdot \left(-\dfrac{2}{5}\right) = -\dfrac{8}{25}$
	The answer makes sense, because 2 1/2 does not fit into 4/5, not even half-way.

7. Divide.

a. $-\dfrac{2}{9} \div \dfrac{6}{7}$

b. $\dfrac{9}{8} \div \left(-1\dfrac{1}{2}\right)$

c. $-10 \div \dfrac{5}{6}$

d. $-\dfrac{1}{9} \div \left(-\dfrac{1}{3}\right)$

e. $10\dfrac{1}{5} \div \left(-2\dfrac{1}{3}\right)$

f. $10 \div \dfrac{1}{6}$

Multiply and Divide Rational Numbers 2

We can **write rational numbers in four different ways**: as percentages, as decimals, as ratios, and as fractions. You need to bear in mind that a calculation done in fractions will look very different from the same calculation done in decimals, yet the answers are equal.

Example 1. Let's do the problem $2.3 \div 8$ using both fraction division and long division.

For fraction division, we need to write 2.3 as 23/10. Compare the two calculations on the right.

The answers – 0.2875 and 23/80 – look very different!

Yet, they are both correct and equal: $23/80 = 0.2875$. (You can check that by converting 23/80 into a decimal with a calculator.)

They look different because the fraction 23/80 is *not* expressed in 10,000th parts like the decimal 0.2875 is. As a fraction, the decimal 0.2875 is 2875/10,000, and that fraction simplifies to 23/80. And that's another way to see that they're equal.

Decimal division:

$$8 \overline{)2.3000} \quad 0.2875$$

$$\begin{array}{r} 0.2875 \\ 8\overline{)2.3000} \\ \underline{16} \\ 70 \\ \underline{-64} \\ 60 \\ \underline{-56} \\ 40 \\ \underline{-40} \\ 0 \end{array}$$

Fraction division:

$$\frac{23}{10} \div 8$$
$$\downarrow \quad \downarrow$$
$$\frac{23}{10} \cdot \frac{1}{8} = \frac{23}{80}$$

Example 2. Let's also look at the division $1.7 \div 0.6$.

To divide decimals, we want a whole-number divisor, so we need to multiply both the dividend and the divisor by 10 to transform the problem into $17 \div 6$.

The decimal division continues on forever, so the answer is the repeating decimal 2.8333... or $2.8\overline{3}$.

The fraction division is very easy to perform because we can simplify before multiplying.

Again, the answers $2.8\overline{3}$ and 2 5/6 look different, but they are equal.

Decimal division:
$1.7 \div 0.6$
becomes $17 \div 6$.

$$\begin{array}{r} 02.833 \\ 6\overline{)17.000} \\ \underline{12} \\ 50 \\ \underline{-48} \\ 20 \\ \underline{-18} \\ 20 \\ \underline{-18} \\ 2 \end{array}$$

Fraction division:

$$\frac{17}{10} \div \frac{6}{10}$$
$$\downarrow \quad \downarrow$$
$$\frac{17}{\cancel{10}} \cdot \frac{\cancel{10}}{6} = \frac{17}{6} = 2\frac{5}{6}$$

(Notice that we ended up making the same transformation in this fraction division that we did in the decimal division: $17 \div 6 = 17/6$.)

Example 3. Let's multiply $2.4 \cdot 0.9$ in both ways.

The decimal multiplication gives us 2.16.

In the fraction multiplication, we first simplify 24/10 to 12/5. Then we simplify one more time before multiplying. The answer is 2 4/25.

Again, the answers look different, but they are equal. Can you see that the fraction 4/25 is equal to 16/100?

Decimal multiplication:

$$\begin{array}{r} 3 \\ 2.4 \\ \underline{\cdot\ 0.9} \\ 2.16 \end{array}$$

Fraction multiplication:

$$\frac{24}{10} \cdot \frac{9}{10}$$
$$\downarrow \quad \downarrow$$
$$\frac{\cancel{12}}{5} \cdot \frac{9}{\cancel{10}} = \frac{54}{25}$$
$$5$$
$$= 2\frac{4}{25}$$

1. Solve using both decimal division and fraction division.

a. $0.5 \div 4$ **Decimal division:** **Fraction division:**

b. $1.3 \div 0.3$ **Decimal division:** **Fraction division:**

c. $0.57 \div 8$ **Decimal division:** **Fraction division:**

2. Solve using both decimal multiplication and fraction multiplication.

a. $0.2 \cdot 0.03$	**Decimal multiplication:**	**Fraction multiplication:**

b. $0.6 \cdot 0.12$	**Decimal multiplication:**	**Fraction multiplication:**

c. $2.2 \cdot 0.75$	**Decimal multiplication:**	**Fraction multiplication:**

If some numbers in your calculation are decimals and others are fractions, these strategies will help:

- Convert the decimals into fractions. Then calculate using fraction arithmetic.
- Treat the fractions as divisions. Then calculate the resulting quotients using decimal arithmetic.

Example 4. Find 56% of 3/7. Also estimate the result with mental math.

Estimate: Since 56% is a little *more* than one-half, and 3/7 is a little *less* than one-half, then 56% of 3/7 should be close to 1/2 of 1/2, or 1/4.

Calculation: Calculating a percentage of a number is easy using decimal multiplication: We write 56% as 0.56. The word "of" corresponds to a multiplication sign. So 56% of 3/7 becomes 0.56 · 3/7.

Think of the fraction 3/7 as a division, and the overall expression equals $\frac{0.56 \cdot 3}{7}$.

It is easy to calculate its value with a calculator or with pencil and paper.

We get $\frac{0.56 \cdot 3}{7} = 0.24$, which is very close to our mental-math estimate of 1/4.

3. Solve *without* a calculator. First change each percentage into a decimal. Then multiply.

a. 24% of $0.30	**b.** 88% of −4	**c.** 60% of −$2,100

4. Solve *without* a calculator. Change decimals into fractions or treat fractions as divisions, whichever is easier.

a. $-\frac{1}{6} \cdot 0.9$	**b.** $\frac{2}{5} \cdot (-1.2)$	**c.** $0.2 \cdot \frac{5}{7}$

5. A ladder that is 2.1 m tall sits in water so that 2/3 of its height is below water. How tall is the part sticking out of the water?

6. A certain town had a budget of $4,500,000, and 40% of it was used to pay the salaries of the town's employees. How many dollars were *not* used for that purpose?

> Which expressions match the problem?
>
> *Margaret sold 3/5 of the 250 candles she had made. How many did she sell?*
>
> $\dfrac{250}{5} \cdot 3 \qquad \dfrac{3 \cdot 250}{5} \qquad 250 \cdot \dfrac{3}{5}$
>
> Don't rush ahead. Think about it.

> The fact is, it does not matter whether you multiply or divide first, as long as the expression doesn't contain any other operations. Therefore, you can first divide 250 by 5, then multiply by 3, *or* first multiply 250 · 3 and then divide by 5. So all of the expressions are equal.
>
> The important thing is that you <u>divide by 5</u>: the 5 is always in the denominator!

7. Write equivalent expressions. You do not have to calculate their value.

   a. $\dfrac{4}{100} \cdot 8$ and $\dfrac{\square \cdot \square}{\square}$ and $\square \cdot \dfrac{\square}{\square}$ b. $\dfrac{5 \cdot 0.4}{8}$ and $\dfrac{\square}{\square} \cdot \square$ c. $40 \cdot \dfrac{5}{8}$ and $\dfrac{\square \cdot \square}{\square}$

8. Match each problem below with one of the expressions from question 7. Solve the problems.

   a. Two students divided $40 unequally so that one of them got 5/8 of it and the other got the rest. How much did the first student get?

   b. The road crew covered 4% of the 8 km stretch on Monday. What distance did they cover on Monday?

   c. A hundred years ago, 5/8 of the factory's workers were immigrants. Of them, 40% were female. What percentage of the factory workers were female immigrants?

9. Solve with a calculator. Round your answer to three decimal digits. Check that your answer makes sense by estimating with mental math.

a. $\dfrac{6}{21} \cdot (-1.095)$	b. $6.97 \cdot \left(-1\dfrac{3}{4}\right)$	c. $\left(-\dfrac{45}{100}\right) \cdot (-78.85)$
d. 80% of $\dfrac{7}{9}$	e. 14% of $1\dfrac{3}{8}$	f. $-\dfrac{23}{49} \cdot \dfrac{49}{80}$

10. Give a real-life context for each multiplication. Then solve. I have already done the first two for you.

    Hint: The area of a rectangle, the length resulting from stretching or shrinking a dimension, a fractional part, and a percentage of a quantity are all calculated by multiplying.

a. $1.28 \cdot 250$ Marsha drew a square on the computer with sides 250 pixels long. Then she stretched it so that the sides became 128% of the sides of the original square. How long are the sides now?   (solve the problem)
b. $(3/5) \cdot 4.30$ A toy that costs \$4.30 is discounted by 2/5 of its price. What is the new price?   (solve the problem)
c. $(9/10) \cdot 2{,}100$ m
d. $0.65 \cdot 19.90$
e. $(2/3) \cdot (3\ 1/2)$
f. $0.9 \cdot 0.2$
g. $(1/2) \cdot 1.6$

11. Give a real-life context for each division. Then solve. Give your answers as fractions or rounded to three decimals. *Hints:* Interpret a decimal number as a money amount. Division by a whole number can represent equal sharing. Division by a fraction or a decimal can represent how many servings or pieces of that size you get out of a certain quantity.

a. 7.28 ÷ 4

b. (3/5) ÷ 0.04

c. 2 3/4 ÷ 8

d. 9 ÷ 1.2

Many Operations with Rational Numbers

In this lesson we practice calculations with rational numbers that involve several operations.

The order of operations still applies. Do you remember the acronym PEMDAS?

1. **P**arentheses
2. **E**xponents
3. **M**ultiplication and **D**ivision, from left to right.
4. **A**ddition and **S**ubtraction, from left to right.

Example 1. Solve $0.4 \div \left(\dfrac{1}{8} + \dfrac{1}{3}\right)$.

Add what is in the parentheses first: $\dfrac{1}{8} + \dfrac{1}{3} = \dfrac{3}{24} + \dfrac{8}{24} = \dfrac{11}{24}$.

It is easier to convert the decimal 0.4 into the fraction 2/5 than to convert the fraction 11/24 into a decimal.

The original problem now becomes $\dfrac{2}{5} \div \dfrac{11}{24} = \dfrac{2}{5} \cdot \dfrac{24}{11} = \dfrac{48}{55}$.

Example 2. Solve $\dfrac{1}{6} + \dfrac{2}{3} - 1\dfrac{3}{4}$.

You can solve this problem in two different ways:

Either: First solve 1/6 + 2/3. Then subtract 1 3/4 from the answer you get.

Or: First find a common denominator for all three fractions (12 works). Convert each of them into an equivalent fraction with that denominator. Then calculate.

In two calculations:

$\dfrac{1}{6} + \dfrac{2}{3} = \dfrac{1}{6} + \dfrac{4}{6} = \dfrac{5}{6}$

$\dfrac{5}{6} - 1\dfrac{3}{4} = \dfrac{10}{12} - \dfrac{21}{12} = -\dfrac{11}{12}$

Converting all the fractions to a common denominator:

$\dfrac{1}{6} + \dfrac{2}{3} - 1\dfrac{3}{4} = \dfrac{2}{12} + \dfrac{8}{12} - \dfrac{21}{12} = -\dfrac{11}{12}$

Example 3. Find the value of $2.95 \div \dfrac{3}{8}$.

Since a fraction is a division problem, we can rewrite this as 2.95 ÷ (3 ÷ 8). However, that is NOT the same as 2.95 ÷ 3 ÷ 8. Please calculate the value of both expressions using your calculator to verify this.

<u>Note:</u> If your calculator does not have parentheses, you need to calculate the value of 3 ÷ 8 first, enter that into the calculator's memory (or write it on paper), and then calculate the rest.

Another way to solve this is to use fraction division, and write it as $\dfrac{295}{100} \div \dfrac{3}{8}$. Then change it into a multiplication by the reciprocal of 3/8 to get $\dfrac{295}{100} \cdot \dfrac{8}{3} = \dfrac{295 \cdot 8}{100 \cdot 3}$. Then you can use a calculator.

<u>Note:</u> You need to enter this into your calculator as 295 · 8 ÷ 300, and *not* as 295 · 8 ÷ 100 · 3. Why? Because any expression in the denominator is really inside (implied) parentheses.

1. Solve.

a. $\dfrac{1}{8} + \dfrac{1}{2} - \dfrac{2}{3}$

b. $1\dfrac{3}{4} + \dfrac{1}{2} \cdot \dfrac{2}{7}$

2. Choose the correct way to enter these calculations into a calculator and solve the expressions.

a. Calculation: $\dfrac{3}{7} \div 1.012$

1. $3 \div 7 \div 1.012$
2. $3 \div (7 \div 1.012)$
3. It does not matter; both will give the correct answer.

b. Calculation: $6.5 \div \dfrac{9}{11}$

1. $6.5 \div 9 \div 11$
2. $6.5 \div (9 \div 11)$
3. It does not matter; both will give the correct answer.

c. Calculation: $0.98 \cdot \dfrac{56}{100}$

1. $0.98 \cdot 56 \div 100$
2. $0.98 \cdot (56 \div 100)$
3. It does not matter; both will give the correct answer.

3. Solve using *fraction* arithmetic and the correct order of operations.

a. $6\dfrac{5}{12} - 2\dfrac{7}{8} - 0.5$

b. $5.6 - \dfrac{5}{6} \cdot 0.9$

c. $\dfrac{3}{8} \div \dfrac{6}{7} - \dfrac{2}{3} \cdot \dfrac{3}{8}$

A **complex fraction** is a fraction that contains another fraction in the numerator, denominator, or both.

For example, $\dfrac{\frac{1}{2}}{\frac{3}{10}}$ is a complex fraction.

We simplify it by treating the MAIN fraction line as a division. We get:

$$\frac{1}{2} \div \frac{3}{10} = \frac{1}{2} \cdot \frac{10}{3} = \frac{10}{6} = \frac{5}{3} = 1\frac{2}{3}.$$

Example 4. Simplify $\dfrac{6}{\frac{2}{5}}$.

This time only the denominator has a fraction. The numerator is a whole number. Using the division symbol, we can write this as

$$6 \div \frac{2}{5} = 6 \cdot \frac{5}{2} = 15.$$

4. Simplify these complex fractions.

a. $\dfrac{4}{\frac{3}{7}}$	b. $\dfrac{\frac{3}{4}}{\frac{2}{9}}$	c. $\dfrac{\frac{5}{6}}{\frac{1}{3}}$
d. $\dfrac{\frac{5}{12}}{7}$	e. $\dfrac{\frac{7}{9}}{10}$	f. $\dfrac{11}{\frac{6}{5}}$

5. Calculate the value of these expressions both with a calculator and using pencil and paper.

a. $\dfrac{12}{-1\frac{1}{2}}$	b. $\dfrac{-\frac{9}{12}}{7.5}$	c. $\dfrac{2.95}{\frac{3}{4}}$

> Alex received 1/3 of a profit of $816, and his business partners got the rest. Alex allotted 3/4 of what he received toward car repairs. So how much money does Alex have to use for car repairs?
>
> This situation involves **a fractional part of a fractional part of a quantity.** We can therefore write the expression $\frac{3}{4} \cdot \frac{1}{3} \cdot \$816$ for the money he uses for car repairs.

6. Find the answer to the problem above. *Hint: you can simplify before you multiply.*

7. The Smiths and the Cars equally shared the cost of putting a fence between their properties. The fence cost $436, and the Smith's children paid 1/5 of their parents' share.

   a. Write and solve a single expression to represent the amount the Smiths' children paid.

   b. Write and solve a single expression to represent the amount Mr. and Mrs. Smith paid.

8. Find the value of the expressions. Give your answer rounded to three decimal digits.

a. $\frac{6}{21} \cdot (-0.095)$	**b.** $9.99 \cdot \frac{3}{8} \cdot 0.5$	**c.** $0.21 \cdot \frac{16}{59} \cdot \frac{28}{10}$

9. Mark was monitoring the temperature on his yard during one winter day. The highest temperature was −2°C and the lowest temperature was −11°C.

   Use the formula F = (9/5)C + 32 to change these two temperatures from Celsius to Fahrenheit degrees. Give your answers rounded to one decimal digit.

Scientific Notation

Do you remember the powers of ten?

The expressions 10^4, 10^{11}, 10^7, and so on are called **powers of ten**. When you have ten to any power, the exponent tells you how many *zeros* to write after the one.

Remember also:

$2 \cdot 10^5$ means $2 \cdot 100{,}000$, which equals $200{,}000$.

$8 \cdot 10^7$ means $8 \cdot 10{,}000{,}000$, which equals $80{,}000{,}000$.

10^6	1,000,000
10^5	100,000
10^4	10,000
10^3	1,000
10^2	100
10^1	10
10^0	1

We can write *any* number as the product of a decimal number between 1 and 10 and a power of ten. This way of writing numbers is called **scientific notation**. The numbers below are written both in scientific notation and in common notation.

Scientific Notation	(in-between calculation)	Common notation
$6.7 \cdot 10^4$	$6.7 \cdot 10{,}000$	67,000
$2.83 \cdot 10^6$	$2.83 \cdot 1{,}000{,}000$	2,830,000
$5.089 \cdot 10^5$	$5.089 \cdot 100{,}000$	508,900
$1.03 \cdot 10^8$	$1.03 \cdot 100{,}000{,}000$	103,000,000

Example 1. How do you write $5.089 \cdot 100{,}000$ in common notation?

A hundred thousand is the largest place value in the number. So simply write the digits "5089" and add enough zeros to put the "5" into the hundred thousands place. The answer is 508,900.

Example 2. How do you write $2.83 \cdot 1{,}000{,}000$ in common notation?

Just write the digits "283" and add enough zeros on the end to put the "2" in the millions place. So $2.83 \cdot 1{,}000{,}000$ becomes $2{,}830{,}000$.

1. Fill in the table with the rest of the numbers written in the indicated ways.

Scientific Notation	(in-between calculation)	Common notation
$6 \cdot 10^5$		
$2.5 \cdot 10^5$		
$5.39 \cdot 10^4$		
$2.03 \cdot 10^6$		
$8.904 \cdot 10^3$		
$1.5594 \cdot 10^8$		

> **Example 3.** Write 25,600 in scientific notation.
>
> First note the largest place value is ten thousands. This gives you the power of ten to use: ten thousand is 10^4. Next, write the digits of 25,600, excluding the trailing zeros, and put a decimal point after the first digit. We get 2.56. So $25,600 = 2.56 \cdot 10^4$.

> **Example 4.** Write 6,078,500,000 in scientific notation.
>
> The largest place value is billions or 10^9. When we exclude the trailing zeros, the digits are 60785. We put a decimal point right after 6 to get 6.0785. So 6,078,500,000 is $6.0785 \cdot 10^9$.

2. Write the numbers with scientific notation.

   a. 13,000

   b. 204,000

   c. 35,600

   d. 4,506,000

   e. 13,080,000

   f. 10,050

   g. 8,300

   h. 289,000

   i. 405,100,000

   j. 4,980,000,000

3. Complete the chart by rewriting the distance in scientific notation.

Planet	Average distance from sun (km)	In scientific notation (km)
Mercury	58,000,000	
Venus	108,000,000	
Mars	227,900,000	
Jupiter	778,570,000	
Uranus	2,870,000,000	
Neptune	4,495,000,000	

4. **Famous Bridges.** Translate the words "million" or "billion" into numerical form to complete the table by writing the costs of the bridges in common notation.

Bridge	Year	Cost (dollars)	Cost in dollars
Golden Gate Bridge	1937	$27 million	$27,000,000
San Diego-Coronado Bridge	1969	$47.6 million	
Erasmusburg Bridge	1996	$110 million	
Oresund Bridge	2000	$3.8 billion	
Strait of Messina Bridge	canceled	$8.6 billion	

Example 5. The number $0.86 \cdot 10^4$ is not in correct scientific notation, because the factor 0.86 should be at least 1 and less than 10. To fix that, we could first write $0.86 \cdot 10^4$ in common notation (by moving the decimal point four places to the right): $0.86 \cdot 10^4 = 0.86 \cdot 10{,}000 = 8{,}600$.

Since the largest place value in 8,600 is thousands, we write $8.6 \cdot 10^3$.

You might have noticed a little shortcut for changing $0.86 \cdot 10^4$ into $8.6 \cdot 10^3$. Since changing 0.86 to 8.6 made it ten times larger, the 10^4 had to be made ten times smaller (10^3).

Example 6. Write the number $210 \cdot 10^5$ correctly in scientific notation.

Instead of 210, we need to use 2.1. We are *dividing* 210 by a hundred. So the 10^5 has to be multiplied by a hundred to get 10^7, and $210 \cdot 10^5 = 2.1 \cdot 10^7$.

If you find the shortcut confusing, just ignore it and convert $210 \cdot 10^5$ into common notation (21,000,000) first, and then change that into scientific notation.

5. Rewrite the numbers into correct scientific notation.

   a. $26 \cdot 10^6$

   b. $0.9 \cdot 10^5$

   c. $358 \cdot 10^4$

   d. $0.208 \cdot 10^7$

   e. $0.02 \cdot 10^8$

   f. $10.1 \cdot 10^6$

6. A 70-kg male body contains approximately 7,000,000,000,000,000,000,000,000,000 atoms. Of these, 20,000,000,000,000,000,000,000 are gold atoms. Write these numbers using scientific notation.

   > Of all these atoms, approximately $4.22 \cdot 10^{27}$ are hydrogen atoms, $1.61 \cdot 10^{27}$ are oxygen atoms, $8.03 \cdot 10^{26}$ are carbon atoms, and $3.9 \cdot 10^{25}$ are nitrogen atoms. See a complete list of elements at https://en.wikipedia.org/wiki/Composition_of_the_human_body

7. Complete the "cross-number puzzle" by writing the numbers in common notation.

   Across:

   d. $2 \cdot 10^3$

   e. $3.1 \cdot 10^2$

   g. $1.003 \cdot 10^6$

   h. $2.99 \cdot 10^7$

   Down:

   a. $8.93 \cdot 10^5$

   b. $4.209 \cdot 10^4$

   c. $7.403 \cdot 10^7$

   f. $5.3 \cdot 10^4$

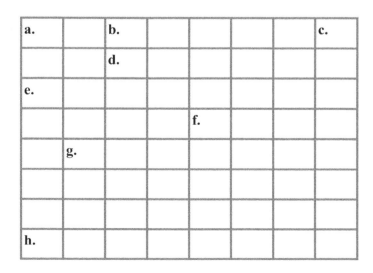

Equations with Fractions

Solve equations with fractions in the same way that you would equations with integers: Just apply the same operation to both sides of the equation to isolate the unknown.

Sometimes we can make things even easier by choosing an operation that turns the fractions into integers.

Example 1. Solve the equation $x - \dfrac{7}{8} = \dfrac{11}{12}$.

We will simply add 7/8 to both sides. We get:

$$x - \frac{7}{8} = \frac{11}{12} \qquad \Big| + 7/8$$

$$x = \frac{11}{12} + \frac{7}{8}$$

$$x = \frac{22}{24} + \frac{21}{24} = 1\frac{19}{24}$$

Lastly, we check that 1 19/24 satisfies the equation:

$$1\frac{19}{24} - \frac{7}{8} \stackrel{?}{=} \frac{11}{12}$$

$$1\frac{19}{24} - \frac{21}{24} \stackrel{?}{=} \frac{11}{12}$$

$$\frac{22}{24} = \frac{11}{12} \checkmark$$

Example 2. Solve the equation $\dfrac{x}{3} = \dfrac{7}{8}$.

Don't get "shook up" by the fraction. Looking at the left side, we see the variable is divided by 3. To isolate it, we simply multiply both sides by 3.

$$\frac{x}{3} = \frac{7}{8} \qquad \Big| \cdot 3$$

$$\frac{3x}{3} = \frac{7}{8} \cdot 3$$

$$x = \frac{21}{8} = 2\frac{5}{8}$$

Lastly, we check that 2 5/8 is indeed a solution by substituting it into the equation in place of x:

$$2\frac{5}{8} \div 3 \stackrel{?}{=} \frac{7}{8}$$

$$\frac{21}{8} \div 3 = \frac{7}{8} \checkmark$$

1. Solve the equations.

a. $x + \dfrac{1}{2} = \dfrac{5}{6}$

b. $x - \dfrac{4}{7} = \dfrac{2}{3}$

2. Solve.

a. $s - \dfrac{7}{2} = 2\dfrac{1}{3}$

b. $1\dfrac{1}{5} + v = \dfrac{3}{10}$

c. $8x = -\dfrac{3}{4}$

d. $\dfrac{z}{8} = -\dfrac{11}{12}$

e. $2\dfrac{3}{8} - x = \dfrac{1}{2}$

f. $\dfrac{2}{11} - x = 3$

3. Three families shared the cost of purchasing a $5,000 generator. The Martins paid 1/3 of the cost, the Millers paid 40%, and the Browns paid the rest.

   a. Find how much the Browns paid.

   b. Check that your answer is reasonable by using mental math and estimation. Explain your reasoning.
      (You *always* need to check that your answers are reasonable!)

4. Write an equation for each problem, and solve. You can also solve the problems using another method just to check that you get the same answer.

a. The perimeter of a square is 14 1/2 inches. How long is one side?	b. The area of a rectangle is 32 1/2 square feet, and one of its sides measures 6 feet. How long is the other side?

Example 3. Solve the equation $\frac{4}{5}x = \$15.88$

The equation says that four-fifths of an unknown is $15.88. Maybe some item was discounted by 1/5 so that it costs only 4/5 of what it did before, and now the discounted price is $15.88.

Notice that the unknown is multiplied by 4/5. So <u>one way</u> to solve it is simply to divide both sides by 4/5 or by 0.8.

<u>Another way</u> is to "undo" the 4/5 in two steps: multiply both sides by 5 (since x is divided by 5) and then divide both sides by 4 (since x is multiplied by 4).

Some other ways are to use logical reasoning or to draw a bar model.

You can "undo" the 4/5 in two steps:

$$\frac{4x}{5} = \$15.88 \quad | \cdot 5$$

$$\frac{5 \cdot 4x}{5} = \$15.88 \cdot 5$$

$$4x = \$79.40 \quad | \div 4$$

$$x = \$19.85$$

5. Solve.

a. $\frac{2}{3}x = 210$

b. $\frac{3}{8}y = \$14.94$

6. Write an equation for each problem, and solve it. You can also solve the problems using some other method just to check that you get the same answer.

a. Three-fifths of a number is 45. What is the number?

b. Two-sevenths of a number is 4.5. What is the number?

7. A cookie recipe calls for 2 1/4 cups of flour. Amanda is substituting 1/3 of the total amount of flour with coconut flour. She is also making a double batch.

   a. How much regular flour and how much coconut flour should Amanda use?

   b. Check that your answer is reasonable by using mental math and estimation. Explain your reasoning to your teacher.

8. Here are more equations if you need more practice. Solve them in your notebook or in the space below, and find the answer to the riddle.

O. $\dfrac{t}{6} = \dfrac{6}{9}$	**C.** $5 - x = 6\dfrac{2}{9}$	**E.** $\dfrac{5}{9}x = 45$	**N.** $x + \dfrac{1}{9} = -\dfrac{1}{9}$
R. $3x = -\dfrac{1}{3}$	**E.** $\dfrac{w}{9} = -\dfrac{1}{2}$	**T.** $-3x = \dfrac{9}{2}$	**M.** $\dfrac{2}{9}x = 6$

Why didn't the quarter go rolling down the hill with the nickel?

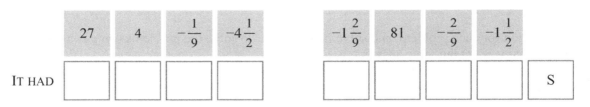

IT HAD ☐ ☐ ☐ ☐   ☐ ☐ ☐ ☐ S

Equations with Decimals

Solve equations with decimals the same way that you solve equations with integers: Apply the same operation to both sides to isolate the unknown.

Example 1. Solve $8.2x = 0.94$.

To isolate x, divide both sides by 8.2.

$$8.2x = 0.94 \quad | \div 8.2$$

$$\frac{8.2x}{8.2} = \frac{0.94}{8.2}$$

$$x \approx 0.1146$$

You can use a calculator to check that $8.2 \cdot 0.1146 = 0.93972$ is close to 0.94. It's not exactly 0.94 because 0.1146 is a rounded answer.

Example 2. Solve $y - 9.85 = -8.7$.

To isolate y, we need to add 9.85 to both sides.

$$y - 9.85 = -8.7 \quad | + 9.85$$

$$y = -8.7 + 9.85$$

$$y = 9.85 - 8.7$$

$$y = 1.15$$

We can change this sum to a subtraction. (Why? Think about an easier problem, $-8 + 9$. It is the same as $9 - 8$.)

Check: $1.15 - 9.85 \stackrel{?}{=} -8.7$

Yes, it checks.

1. Solve the equations. Check your solutions.

a.	$-2.7 = x + 1.5$		b.	$x + 6.1 = -7.5$
c.	$0.92 = z - 0.05$		d.	$y + 0.56 = 0.19$
e.	$y - (-0.5) = 1.24$		f.	$2 - c = -0.61$

2. Practice some more. Solve. Round your answers to three decimal digits.

a.	$\dfrac{x}{2} = -4.48$	b.	$\dfrac{s}{-0.7} = -0.5$
c.	$2.1x = -40$	d.	$24g = 19.38$
e.	$\dfrac{c}{485.3} = -14.3$	f.	$-0.99x = -1.05$

3. Here is another pattern of growth. Draw steps 4 and 5 and answer the questions.

Step   1   2   3

a. How is this pattern growing?

b. How many flowers will there be in step 39?

c. In step n?

d. In which step will the pattern have 8,500 flowers?

Example 3. A set of bed sheets has been discounted by 1/5 and now costs $25.90. What was its original price?

After the discount, 4/5 of the price is left. So 4/5 of the original price is $25.90.

You could solve this problem using logical reasoning (how?) but this time, we will use an equation.

The original price is an unknown — let's denote it by p. From the sentence

4/5 of the original price is $25.90

we can easily write the equation

$$(4/5) \cdot p = \$25.90$$

To solve this equation, you can divide both sides by 4/5. Another, perhaps easier, strategy is to write the fraction 4/5 as 0.8, and then divide both sides of the equation by that.

4. Solve the above problem.

5. Solve each problem using an equation and also using some other strategy, such as a bar model or mental reasoning.

a. Two-fifths of a number is 6.78. What is the number?	
Equation:	Another way:

b. A board game was discounted by 7/10 of its price, so now it costs $11.55. What was the original price?	
Equation:	Another way:

Chapter 4 Mixed Review

1. Jeremy purchased *x* pairs of gloves for $3 each and one pair of rubber boots for $9.

    a. Write an expression for the total cost of his purchases.

    b. The total cost of Jeremy's purchases was $57.
    Write an equation for this situation.

    c. How many pairs of gloves did he buy?
    You can solve the equation or figure it out using mental math.

2. Solve. Simplify one side first.

a. $\quad 2r - 5 \;=\; 10 - (-2)$	**b.** $\quad 2 \cdot 3 \;=\; 9 - 6y$

3. Light travels at a speed of 299,792.458 kilometers per second. That is quite fast!

    a. A beam of light is sent from the earth to the moon. How long will it take to reach the moon?
    The average distance between the earth and the moon is 384,403 km.
    Give your answer to five decimal digits.

    b. (Optional.) Choose some other object in our solar system, and calculate how long it will take a radio message sent from the earth to reach that object. Radio waves travel at the speed of light.
    Use an encyclopedia to find the distances you need to solve the problem.

4. Solve. Check your solutions.

a. $2 - x = -6$	**b.** $-10 - x = 7$
c. $2x = -5$	**d.** $2 + (-11) = 8 + z$

5. Divide and simplify if possible.

a. $1 \div (-3)$	**b.** $-16 \div 20$	**c.** $-45 \div (-36)$

6. Write an addition or subtraction using integers to match the situation.

   a. A shark was swimming at the depth of 4 m. Then it rose 2 m.

   b. Michael owed $250. He then purchased a computer on his credit card for $500 (adding to his debt).

7. Solve. Check your solutions.

a. $\dfrac{x}{-13} = 4$	**b.** $\dfrac{w}{-3} = -11 + 5$
c. $-31 = \dfrac{1}{6}x$	**d.** $1 = -5x$

Chapter 4 Review

1. Write these numbers as a ratio (fraction) of two integers.

| a. −3 | b. 30 | c. 0.21 | d. −1.9 |

2. Mark the decimals on the number line: −0.21, −0.7, −0.03, −0.92

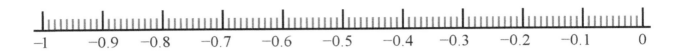

3. Write these decimals as fractions.

| a. 0.0472 | b. −1.02938442 | c. 2.38166 |

4. Write the fractions as decimals.

| a. $-\dfrac{24}{10{,}000}$ | b. $\dfrac{9{,}872}{10}$ | c. $\dfrac{4{,}593}{100{,}000}$ |

5. Write these repeating decimals using a horizontal line over the repeating part.

   a. 0.21212121...
   b. 1.099555555...

6. Write these repeating decimals using an ellipsis (three periods).

   a. $2.06\overline{9}$
   b. $0.006\overline{812}$

7. Which is more, 0.7, or $0.\overline{7}$?
   How much more?

8. Are all terminating decimals rational numbers?

   If not, give an example of a terminating decimal that is not a rational number.

9. Are all repeating decimals rational numbers?

   If not, give an example of a repeating decimal that is not a rational number.

10. Write as decimals. Calculate each answer to at least six decimal places. If you find a repeating pattern, then indicate the repeating part. If you don't, then round your answer to five decimals.

a. $\dfrac{3}{22}$	b. $1\dfrac{14}{23}$

11. The distance between two numbers a and b is given by the expression $|a - b|$. Show that this expression indeed gives the distance between the two given values of a and b by evaluating the expression and by also calculating the distance using logical reasoning.

a. $a = 6$ and $b = -7$	b. $a = -1.3$ and $b = -7.6$
Distance:	Distance:
Absolute value of the difference:	Absolute value of the difference:
$| - | =$	$| - | =$

12. Multiply mentally.

a. $0.2 \cdot 0.07$	b. $-0.8 \cdot 0.005$	c. $(-0.2)^3$
d. $-5 \cdot (-2.2)$	e. $-0.2 \cdot 0.1 \cdot (-0.3)$	

13. Multiply

a. $-\dfrac{4}{11} \cdot \left(-\dfrac{7}{12}\right)$	b. $\dfrac{5}{6} \cdot \left(-3\dfrac{1}{2}\right)$	c. $-\dfrac{9}{20} \cdot \dfrac{2}{3} \cdot \left(-\dfrac{1}{5}\right)$

14. Divide.

a. $1\dfrac{1}{5} \div \left(-\dfrac{1}{4}\right)$	b. $21 \div 0.06$

15. Solve *without* a calculator.

a. 60% of $18	b. $\dfrac{1}{4} \cdot 9.6$	c. $-0.3 \cdot \dfrac{8}{11}$

16. Simplify these complex fractions.

a. $\dfrac{3}{\frac{2}{5}}$	b. $\dfrac{\frac{6}{7}}{\frac{5}{12}}$	c. $\dfrac{\frac{8}{3}}{2}$

17. Give a real-life context for each calculation. Then solve.

   a. $1.56 \cdot 0.8$

   b. $6 \div (1/2)$

18. A **kilowatt-hour** (kWh) is a unit of energy. It describes the energy consumed (by some device) that uses one **kilowatt** (kW) of power for one hour. The formula is: **power × time = energy**. So you multiply the kilowatts times the hours to get kilowatt-hours. For example, let's say a 2kW air conditioner runs for one hour. Then it uses 2 kWh (two kilowatt-hours) of energy in that time. Your electric company charges you for the amount of *energy* that you consume.

   If electricity costs 16.86 cents per kWh, then how much would it cost to run a 2kW air conditioner for 16 hours each day during June, July, and August?
   Hint: First calculate how much energy (in kilowatt-hours) the AC unit uses in that time period.

19. Write the numbers in scientific notation.

   a. 6,798,000

   b. 56,000,000,000

20. Write the numbers in numerical form.

   a. $7.8 \cdot 10^5$

   b. $3.4958 \cdot 10^9$

21. Solve the problem using an equation and also using some other strategy.

A forest fire was 7/10 contained. The contained area was 4,200 acres. What was the total area of the fire?	
Equation:	Another way:

22. Solve.

a. $x - \dfrac{2}{9} = 5\dfrac{1}{20}$	**b.** $5y = -\dfrac{4}{12}$
c. $0.94 = 1.1 - x$	**d.** $-0.3x = 10$

Chapter 5: Equations and Inequalities
Introduction

In this chapter we delve deeper into our study of equations. Now, the equations require two or more steps to solve and may contain parentheses. The variable may appear on both sides of the equation. Students will also write equations to solve simple word problems.

There is also another lesson on patterns of growth, which may seem to be simply a fascinating topic, but in reality presents the fundamentals of a very important concept in algebra—that of linear functions (although they are not mentioned by that name)—and complements the study of lines in the subsequent lessons.

After the section about equations, the text briefly presents the basics of inequalities and how to graph them on a number line. Students apply the principles for solving equations to solve simple inequalities and word problems that involve inequalities.

The last major topic is graphing. Students begin the section by learning to graph linear equations and continue on to the concept of slope, which in informal terms is a measure of the inclination of a line. More formally, slope can be defined as the ratio of the change in *y*-values to the change in *x*-values. The final lesson applies graphing to the previously-studied concepts of speed, time, and distance through graphs of the equation $d = vt$ in the coordinate plane.

You might consider mixing the lessons from this chapter with lessons from some chapter from part B (such as chapter 6 or 8). For example, the student could study topics from this chapter and from the geometry chapter on alternate days, or study a little from each chapter each day. Such, somewhat spiral, usage of the curriculum can help prevent boredom, and also to help students retain the concepts better.

As a reminder, you will find free videos covering many topics of this chapter of the curriculum at **https://www.mathmammoth.com/videos/** (choose 7th grade).

The Lessons in Chapter 5

	page	span
Two-Step Equations	168	*5 pages*
Two-Step Equations: Practice	173	*4 pages*
Growing Patterns 2	177	*4 pages*
A Variable on Both Sides	181	*6 pages*
Some Problem Solving	187	*3 pages*
Using the Distributive Property	190	*6 pages*
Word Problems	196	*3 pages*
Inequalities	199	*5 pages*
Word Problems and Inequalities	204	*2 pages*
Graphing	206	*4 pages*
An Introduction to Slope	210	*5 pages*
Speed, Time and Distance	215	*5 pages*
Chapter 5 Mixed Review	220	*3 pages*
Chapter 5 Review	223	*6 pages*

Helpful Resources on the Internet

You can also access this list of links at https://links.mathmammoth.com/gr7ch5

Pre-algebra Videos
A set of videos by the author about equations, matching this pre-algebra course.
https://cutt.ly/Pre-Algebra-Videos-by-Maria

TWO-STEP EQUATIONS

Balance Beam Equations
Represent the given equations using draggable objects. Then you can "solve" the equation by removing the same amounts from both sides, until one x is alone on one side.
https://www.geogebra.org/m/MG7eZX3g#material/mMGMgTYb

Two-Step Equations Game
Choose the correct root for the given equation (multiple-choice), and then you get to attempt to shoot a basket. The game can be played alone or with another student.
https://bit.ly/Two-Step-Equations-Game

Two-Step Equations
Practice solving equations that take two steps to solve in this interactive exercise from Khan Academy.
https://bit.ly/Two-Step-Equations

Two-Step Equations Word Problems (from Khan Academy)
Practice writing equations to model and solve real-world situations in this interactive exercise.
https://bit.ly/two-step-equations-word-problems

Visual Patterns
Hundreds of growing patterns. The site provides the answer to how many elements are in step 43 of the pattern.
https://www.visualpatterns.org/

Matchstick Patterns
Describe the sequences of matchstick patterns with a formula.
https://www.transum.org/Maths/Activity/Matchstick_Patterns/

Equations of Sequence Patterns
An instructional video from Khan Academy.
https://www.youtube.com/watch?v=_3BnyEr5fG4

Write Equations for Non-Linear Patterns
Learn how to break down complex shapes that grow in 2 dimensions into smaller terms, making it easy to write an equation and find values for graphing.
https://www.youtube.com/watch?v=ecARQzwvN9w

SIMPLIFYING EXPRESSIONS

Factor the Expressions Quiz
Factor expressions. For example, $-4x + 16$ factors into $-4(x - 4)$.
https://www.thatquiz.org/tq-0/?-jh00-l4-p0

Simplifying Algebraic Expressions Practice Problems
Practice simplifying expressions such as $4(2p - 1) - (p + 5)$ with these 10 questions. Answer key included.
https://www.algebra-class.com/algebraic-expressions.html

Brackets
Expand algebraic expressions containing brackets and simplify the resulting expression.
https://www.transum.org/software/SW/Starter_of_the_day/Students/Brackets.asp?Level=4

Simplifying Algebraic Expressions (1)
Eight practice problems that you can check yourself about combining like terms and using the distributive property.
http://www.algebralab.org/lessons/lesson.aspx?file=Algebra_BasicOpsSimplifying.xml

Collecting Like Terms
Practice your algebraic simplification skills with this interactive online activity.
https://www.transum.org/Maths/Activity/Algebra/Collecting_Like_Terms.asp?Level=1

Expression Exchange Game
Explore how combining algebraic expressions works, and play a game where you combine simple expressions.
https://www.mathmammoth.com/practice/expression-exchange

MORE EQUATIONS

Balance When Adding and Subtracting Game
An interactive balance where you add or subtract x's and 1's until you leave x alone on one side.
https://www.mathsisfun.com/algebra/add-subtract-balance.html

Balancing Equations
An interactive balance shows you an equation. You can then add or subtract the same amount from both sides, or divide both sides by the same number, in order to solve the equation.
https://www.geogebra.org/m/Z5Nzfpqx

Solve Equations Quiz
A 10-question online quiz where you need to solve equations with an unknown on both sides.
https://www.thatquiz.org/tq-0/?-j102-l4-p0

Equations Level 3 Online Exercise
Practice solving equations with an unknown on both sides in this self-check online exercise.
https://www.transum.org/software/SW/Starter_of_the_day/Students/Equations.asp?Level=3

Missing Lengths
Try to figure out the value of the letters used to represent the missing numbers.
https://www.transum.org/software/SW/Starter_of_the_day/Students/Missing_Lengths.asp

Equations Level 4 Online Exercise
Practice solving equations which include brackets in this self-check online exercise.
https://www.transum.org/software/SW/Starter_of_the_day/Students/Equations.asp?Level=4

Equations Level 5 Online Exercise
This exercise includes more complex equations requiring multiple steps to find the solution.
https://www.transum.org/software/SW/Starter_of_the_day/Students/Equations.asp?Level=5

Two-Step Equations Quiz
Practice solving two-step equations with this interactive multiple-choice quiz.
https://www.softschools.com/quizzes/algebra/solving_twostep_equations/quiz9248.html

Rags to Riches Equations
Choose the correct root to a linear equation.
https://www.quia.com/rr/4096.html

Algebra Four Game
To practice the types of equations we study in this chapter, choose "Level 1," and tick the boxes "Variable on both sides," "Distributive Property," and "Two-Step Problems" (don't check "Quadratic Equations").
http://www.shodor.org/interactivate/activities/AlgebraFour/

Solve Equations Exercises
Click "new problem" (down the page) to get a randomly generated equation to solve. This exercise includes an optional graph which the student can use as a visual aid.
https://www.onemathematicalcat.org/algebra_book/online_problems/solve_lin_int.htm#exercises

Equation Word Problems Quiz
Solve the word problems by writing the correct equation, then solving it.
https://reviewgamezone.com/mc/candidate/test/?test_id=22515&title=Equation%20Word%20Problems

Whimsical Windows - Equation Game
Write an equation for the relationship between x and y based on a table of x and y values. Will you discover the long lost black unicorn stallion?
https://mrnussbaum.com/whimsical-windows-online-game

INEQUALITIES

Inequality
A six-question quiz on simple inequalities and their graphs.
https://www.mathopolis.com/questions/q.php?id=12194

Equations and Inequalities Quiz
Test your skills with this multiple-choice quiz.
https://www.proprofs.com/quiz-school/story.php?title=solving-equations-inequalities

Plot Simple Inequalities
Practice plotting simple inequalities on a number line in this 10-question interactive quiz.
https://www.thatquiz.org/tq-o/?-j18-l1-p0

Match Inequalities and Their Plots
Match the statements with the corresponding diagrams in this interactive online activity.
https://www.transum.org/software/SW/Starter_of_the_day/Students/InequalitiesB.asp?Level=5

Solve Simple Inequalities
For each inequality, find the range of values for x which makes the statement true. An example is given.
https://www.transum.org/software/SW/Starter_of_the_day/Students/InequalitiesC.asp?Level=6

Two-Step Inequality Word Problems
Practice constructing, interpreting, and solving linear inequalities that model real-world situations.
https://bit.ly/Two-Step-Inequality-Word-Problems

SPEED, TIME, AND DISTANCE

Representing Motion
A tutorial an interactive quiz with various questions about speed, time, and distance.
https://www.bbc.co.uk/bitesize/guides/z2wy6yc/revision/1

Speed and Velocity
A short illustrated lesson about speed and velocity. There is a quiz at the end of the page, which includes both easier and challenging questions.
https://www.mathsisfun.com/measure/speed-velocity.html

Travel Graphs
Test your understanding of distance-time and speed-time graphs with this self-check exercise.
https://www.transum.org/Maths/Activity/Travel_Graphs/

Distance vs. Time Graphs
This is a series of simulations designed to help students better understand distance-time graphs.
https://www.geogebra.org/m/FYdzG77h

Distance Versus Time Graph Puzzles
Try to move the stick man along a number line in such a way as to illustrate the graph that is shown.
https://davidwees.com/graphgame/

GRAPHING AND SLOPE

Graph Linear Equations
A ten-question online quiz where you click on three points on the coordinate grid to graph the given equation.
https://www.thatquiz.org/tq-0/?-j10g-l4-p0

Find the Slope
A ten-question online quiz that asks for the slope of the given line.
https://www.thatquiz.org/tq-0/?-j300-l4-p0

Slope Slider
Use the sliders to change the slope and the *y*-intercept of a linear equation to see what effect they have on the graph of the line.
http://www.shodor.org/interactivate/activities/SlopeSlider/

Graphing Equations Match
Match the given equations to their corresponding graphs.
http://www.math.com/school/subject2/practice/S2U4L3/S2U4L3Pract.html

Find Slope from Graph
Find the slope of a line on the coordinate plane in this interactive online activity.
https://bit.ly/Find-Slope-from-Graph

Slope - Exercises
Practice finding the slope in this interactive online exercise.
https://www.onemathematicalcat.org/algebra_book/online_problems/compute_slope.htm#exercises

Graphing and Slope Quiz
Practice linear equations and functions in this interactive online quiz.
https://www.thatquiz.org/tq-0/?-j30g-l4-mpnv600-p0

Graph the Line
Drag the points A and B until the line through them matches the given equation.
https://www.geogebra.org/m/NVKV5AFB#material/Hyz9Xuu5

GENERAL

Algebra Quizzes
A variety of online algebra quizzes from MrMaisonet.com.
https://maisonetmath.com/algebra/algebra-quizzes

Two-Step Equations

Just like the name says, **two-step equations take two steps to solve.** We need to apply two different operations to both sides of the equation. Study the examples carefully. It is not difficult at all!

Example 1. Solve $2x + 3 = -5$.

On the side of the unknown (left), there is a multiplication by 2 and an addition of 3. To isolate the unknown, we need to undo those operations.

$$2x + 3 = -5 \quad | -3$$
$$2x = -8 \quad | \div 2$$
$$x = -4$$

Check:
$$2 \cdot (-4) + 3 \stackrel{?}{=} -5$$
$$-8 + 3 \stackrel{?}{=} -5$$
$$-5 = -5 \checkmark$$

What if you divide first?

In this equation you *could* start by dividing by 2 and then subtract next. However, it is easier to subtract first, then divide, because that way you avoid dealing with fractions.

The solution below shows the steps if you divide by 2 first. Notice that the 3 on the left side also has to be divided by 2 to become 3/2.

$$2x + 3 = -5 \quad | \div 2$$
$$x + (3/2) = -5/2 \quad | -3/2$$
$$x = -5/2 - 3/2$$
$$x = -4$$

1. Solve. Check your solutions (as always!).

a. $\quad 5x + 2 = 67$

b. $\quad 3y - 2 = 71$

c. $\quad -2x + 11 = 75$

d. $\quad 8z - 2 = -98$

Example 2. In the equation below, the easiest thing to do is to multiply by 7 first, and then subtract.

$$\frac{x+2}{7} = 12 \quad \Big| \cdot 7$$

$$\frac{7 \cdot (x+2)}{7} = 84$$

$$x + 2 = 84 \quad \Big| -2$$

$$x = 82$$

Check:

$$\frac{82+2}{7} \stackrel{?}{=} 12$$

$$\frac{84}{7} \stackrel{?}{=} 12$$

$$12 = 12 \checkmark$$

2. Solve. Check your solutions (as always!).

a. $\dfrac{x+6}{5} = 14$

b. $\dfrac{x+2}{7} = -1$

c. $\dfrac{x-4}{12} = -3$

d. $\dfrac{x+1}{-5} = -21$

What if, in example 2, you subtract 2 first? (Optional)

It does not work. Subtracting 2 from both sides makes the equation more complicated.

$$\frac{x+2}{7} = 12 \quad \Big| -2$$

$$\frac{x+2}{7} - 2 = 10$$

The quantity $(x + 2)$ is <u>also divided by 7.</u> It is $(x + 2)/7$. Notice that, when calculating the value of this expression, the parentheses indicate that the addition is to be done first, and the division last. Therefore, when *undoing* the operations, we need to undo the division first.

However, it is possible to subtract first. But you need to subtract 2/7 instead of 2. To see that, let's write the expression on the left side in a different way:

$$\frac{x+2}{7} = 12$$

$$\frac{x}{7} + \frac{2}{7} = 12 \quad \Big| -2/7$$

$$\frac{x}{7} = 11\,5/7 \quad \Big| \cdot 7$$

$$x = 82$$

Because it involves fractions, this solution is more complicated than the one shown in Example 2.

Example 3. Again, the unknown is "tangled up" with two different operations (division and addition), so to isolate it, we need two steps.

$$\frac{x}{4} + 5 = -2 \quad \Big| -5$$

$$\frac{x}{4} = -7 \quad \Big| \cdot 4$$

$$x = -28$$

Check:
$$\frac{-28}{4} + 5 \stackrel{?}{=} -2$$
$$-7 + 5 \stackrel{?}{=} -2$$
$$-2 = -2 \checkmark$$

Solving it another way (optional)

In case you wonder if we could multiply by 4 first, yes, in this case we can.

$$\frac{x}{4} + 5 = -2 \quad \Big| \cdot 4$$

$$x + 20 = -8 \quad \Big| -20$$

$$x = -28$$

Note: In the first step, <u>both</u> terms on the left side (x/4 and 5) have to be multiplied by 4!

It is a common student error to multiply only the first term of an expression by a number and to forget to multiply the other terms by that number.

3. Solve. Compare equation (a) to equation (b). They are similar, yet different! Make sure you know how to solve each one.

a. $\dfrac{x}{10} + 3 = -2$

b. $\dfrac{x+3}{10} = -2$

4. Solve. Compare equation (a) to equation (b). They are similar, yet different! Make sure you know how to solve each one.

a. $\dfrac{x}{7} - 8 = -5$

b. $\dfrac{x-8}{7} = -5$

Example 4. What's different about this one?   Check:

$$6 - 2n = 9 \qquad |-6$$
$$-2n = 3 \qquad |\div(-2)$$
$$n = -1\ 1/2$$

$$6 - 2 \cdot (-1\ 1/2) \overset{?}{=} 9$$
$$6 + 3 \overset{?}{=} 9 \checkmark$$

5. Solve. Check your solutions (as always!).

a. $1 - 5x = 2$

b. $12 - 3y = -6$

c. $10 = 8 - 4y$

d. $7 = 5 - 3t$

6. Choose from the expressions at the right to build an equation that has the root $x = 2$.

$2x - 10 \qquad 2x + 10 \qquad 12$
$5x + 6 \qquad 3x - 9$
$14 \qquad \qquad 3 \cdot 3$
$5x - 6$

7. Choose from the expressions at the right to build an equation that has the root $x = 5$.

$1 - 2x \qquad -4 \qquad 3x - 10$
$-8 \cdot 3 \qquad -2 \qquad -9 - 3x$
$-3x + 6 \qquad -2x - 1$

Example 5. Solve as a decimal.

$$\frac{3x}{7} = 0.9 \quad \bigg| \cdot 7$$

$$\frac{7 \cdot 3x}{7} = 6.3$$

$$3x = 6.3 \quad \bigg| \div 3$$

$$x = 2.1$$

Check:
$$\frac{3 \cdot 2.1}{7} \stackrel{?}{=} 0.9$$
$$\frac{6.3}{7} \stackrel{?}{=} 0.9$$
$$0.9 = 0.9 \checkmark$$

In this example, you *could* first divide by 3 and then multiply by 7. The solution wouldn't be any more difficult that way.

8. Solve.

a. $\dfrac{2x}{5} = 8$

b. $\dfrac{3x}{8} = -9$

c. $15 = \dfrac{-3x}{10}$

d. $2 - 4 = \dfrac{s+4}{5}$

e. $\dfrac{x}{2} - (-16) = -5 \cdot 3$

f. $2 - 4p = -0.5$

172

Two-Step Equations: Practice

Example 1. The number line diagram illustrates the equation $-29 + x + 7 = -4$:

Think of starting at -29, jumping x steps, jumping another 7 steps, and arriving at -4.

You can find the value of the unknown x using logical thinking or by writing an equation and solving it. Here is the solution using an equation:

$$-29 + x + 7 = -4 \quad \text{(add } -29 + 7 \text{ on the left side)}$$
$$-22 + x = -4$$
$$+22 \quad +22$$
$$x = 18$$

1. Write an equation to match the number line model and solve for the unknown.

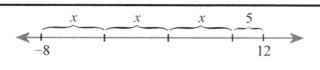

a.

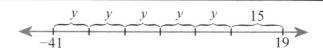

b.

2. Solve. Compare equation (a) to equation (b). They are similar, yet different!

a. $\quad 2 - 5y = -11$

b. $\quad 5y - 2 = -11$

3. Solve. Compare equation (a) to equation (b). They are similar, yet different!

a. $\dfrac{2x}{7} = -5$

b. $\dfrac{x+2}{7} = -5$

4. Solve. Check your solutions (as always!).

a. $20 - 3y = 65$

b. $6z + 5 = -2.2$

c. $\dfrac{t+6}{-2} = -19$

d. $\dfrac{y}{6} - 3 = -0.7$

Example 2. The perimeter of an isosceles triangle is 26 inches and its base measures 5 inches. How long are the two sides that are equal (congruent)?

(To help yourself, label the image with the data from the problem.)

Read the two solutions below. Notice how neatly they tie in with each other!

Solution 1: an equation

Let x be the unknown side length. We get:

$$\text{perimeter} = x + x + 5$$
$$26 = 2x + 5$$

Next we solve the equation:

$$\begin{array}{rl} 2x + 5 = & 26 \quad \big| -5 \\ 2x = & 21 \quad \big| \div 2 \\ x = & 10.5 \end{array}$$

The two other sides measure 10.5 inches each.

Solution 2: Logical thinking/mental math

The perimeter is 26 inches. This means that the two unknown sides and the 5-inch side add up to 26 inches.

Therefore, if we subtract the base side, the two unknown sides must add up to 21 inches.

So one side is half of that, or 10.5 inches.

5. Solve each problem below in two ways: write an equation, and use logical reasoning/mental math.

a. A quadrilateral has three congruent sides. The fourth side measures 1.4 m. If the perimeter of the quadrilateral is 7.1 meters, what is the length of each congruent side?

Equation:	Mental math/logical thinking:

b. You bought six identical baskets from an artisan. She gave you a $12 discount on your order, and your total bill was only $46.80. What is the normal price of one basket?

Equation:

Logical thinking:

c. Two-fifths of a number is 466. What is the number?

Equation:

Logical thinking:

Growing Patterns 2

Step   1   2   3   4   5   6

How do you think this pattern is growing?

How many flowers will there be in step 39?

This pattern adds 2 flowers in each step, except in step 1. This means that by step 39, we have added 2 flowers 38 times. Therefore, there are $1 + 2 \cdot 38 = 77$ flowers in step 39.

Write a formula for the number of flowers in step n.

There are several ways to do this. The three ways explained below are not the only ones!

1. Let's view the pattern as adding 2 flowers in each step after the first one. By step n, the pattern has added one less than n times 2 flowers, because we need to exclude that first step. This means that $(n - 1)$ times 2, or $(n - 1) \cdot 2$, flowers added to the one flower that we started with.

   This gives us the expression $1 + (n - 1) \cdot 2$. Since we customarily put the variable first and the constant last, we can rewrite that expression as $1 + 2(n - 1)$ and then as $2(n - 1) + 1$.

2. Another way to think about this pattern is as two legs. One leg includes the flower in the corner, so it has the same number of flowers as the step number. The other leg doesn't have the corner flower, so it has one flower less than the step numbers. In other words, in step 3, we have $3 + 2$ flowers. In step 4, we have $4 + 3$ flowers. In step 5, we have $5 + 4$ flowers.

   This gives us a formula for the number of flowers in step n: there are $n + (n - 1)$ flowers in step n.

3. Yet another way is that, in each step, there are twice as many flowers as the step number, minus one for the flower that is shared. For example, in step 4, we have twice 4 minus 1, which is seven flowers.

   This also gives us a formula: there are $2n - 1$ flowers in step n.

All of the formulas are equivalent (just as we would expect!) and simply represent different ways of thinking about the number of flowers in each step. On the right, you can see how the first two formulas can be simplified to the third one.	$n + (n - 1)$ $= n + n - 1$ $= 2n - 1$	$2(n - 1) + 1$ $= 2n - 2 + 1$ $= 2n - 1$

In which step are there 583 flowers?

We can use our formula to write an equation to answer this question. In the question the step number n is unknown, but the total number of flowers in that step is 583. Since we know from our formula that there are $2n - 1$ flowers in step n, we get

$$2n - 1 = 583 \quad | +1$$
$$2n = 584 \quad | \div 2$$
$$n = 292$$

1   2   3   4   5

1. **a.** How is this pattern growing?

   b. How many triangles will there be in step 39?

   c. Write a formula for the number of triangles in step *n*.
      Check your answer with your teacher before going on to part (d).

   d. In which step will there be 311 triangles?
      Write an equation and solve it.
      Notice, this question is <u>different</u> from the one in part (c).

1   2   3   4   5

2. **a.** How do you think this pattern is growing?

   b. How many snowflakes will there be in step 39?

   c. Write a formula for the number of snowflakes in step *n*.
      Check your answer with your teacher before going on to part (d).

   d. In which step will there be 301 snowflakes?
      Write an equation and solve it.

> Instead of showing the steps of the pattern horizontally, like this... ...we can also show them like this:
>
>
>
> Now, each row of flowers is one step of the pattern.

3. A section of a flower garden has rows of flowers. The first row has four flowers, and each row after that has one more flower than the previous row.

   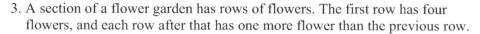

Row	Flowers
1	3 + 1
2	3 + 2
3	3 + 3
4	3 + 4
n	

   a. Write a formula that tells the gardener the number of flowers in row n.

   b. How many flowers are in the 28th row?

   c. In which row will there be 97 flowers? Write an equation and solve it.

4. This pattern is similar to the previous one. This time each row has *two* more flowers than the previous row. Notice that the number of flowers in each row gives us the list of all the odd numbers.

   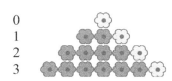

   Here's one way to look at this pattern: Label the first row as "row 0." Then the number of flowers in any row is twice the row number plus 1.

   a. Write a formula that tells the gardener the number of flowers in row n.

   b. In which row will there be 97 flowers? Write an equation and solve it.

5. Each floor of a multi-story building is 9 feet high.

   a. Write an expression for the total height of the building, if it is *n* stories high.

   b. Write an expression for the total height of the building if it is *n* stories high, and the bottom of the first floor is elevated 2 ft above the ground.

   c. How many stories does the building have if the total height of the building is 164 ft? Solve this problem in two ways: using logical thinking and using an equation.

6. Jeremy earns $400 a week. He also earns $15 for every hour he works overtime.

   a. Write an expression for Jeremy's *total* earnings if he works *n* hours overtime. You can use the table to help you.

   b. How many hours should Jeremy work overtime in order to earn $970 in a week?

Overtime hours	Total earnings
0	
1	
2	
17	
n	

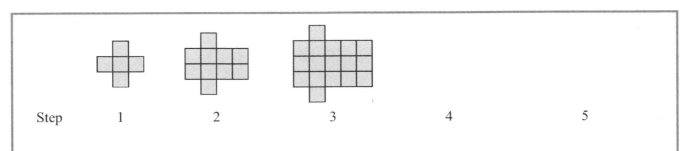

Step   1   2   3   4   5

What is the pattern of growth here?
How many squares will there be in step 59?

Puzzle Corner

A Variable on Both Sides

Example 1. Solve $2x + 8 = -5x$.

Notice that the unknown appears on both sides of the equation. To isolate it, we need to

- either subtract $2x$ from both sides—because that makes $2x$ disappear from the left side
- or add $5x$ to both sides—because that makes $-5x$ disappear from the right side.

$$
\begin{aligned}
2x + 8 &= -5x & &|+5x \\
2x + 8 + 5x &= 0 & &\text{(now add } 2x \text{ and } 5x \text{ on the left side)} \\
7x + 8 &= 0 & &|-8 \\
7x &= -8 & &|\div 7 \\
x &= -8/7
\end{aligned}
$$

Check:
$$
\begin{aligned}
2 \cdot (-8/7) + 8 &\stackrel{?}{=} -5 \cdot (-8/7) \\
-16/7 + 8 &\stackrel{?}{=} 40/7 \\
-2\,2/7 + 8 &= 5\,5/7 \quad \checkmark
\end{aligned}
$$

Example 2. Solve $10 - 2s = 4s + 9$.

To isolate s, we need to

- either add $2s$ to both sides
- or subtract $4s$ from both sides.

The choice is yours. Personally, I like to keep the unknown on the left side and eliminate it from the right.

$$
\begin{aligned}
10 - 2s &= 4s + 9 & &|-4s \\
10 - 2s - 4s &= 9 & &\text{(now simplify } -2s - 4s \text{ on the left side)} \\
10 - 6s &= 9 & &|-10 \\
-6s &= -1 & &|\div(-6) \\
s &= 1/6
\end{aligned}
$$

Check:
$$
\begin{aligned}
10 - 2 \cdot (1/6) &\stackrel{?}{=} 4 \cdot (1/6) + 9 \\
10 - 2/6 &\stackrel{?}{=} 4/6 + 9 \\
9\,4/6 &= 9\,4/6 \quad \checkmark
\end{aligned}
$$

1. Solve. Check your solutions (as always!).

a. $\quad 3x + 2 = 2x - 7$

b. $\quad 9y - 2 = 7y + 5$

2. Solve. Check your solutions (as always!).

a. $\quad 11 - 2q \;=\; 7 - 5q$	**b.** $\quad 6z - 5 \;=\; 9 - 2z$
c. $\quad 8x - 12 \;=\; -1 - 3x$	**d.** $\quad -2y - 6 \;=\; 20 + 6y$
e. $\quad 6w - 6.5 \;=\; 2w - 1$	**f.** $\quad 5g - 5 \;=\; -20 - 2g$

Combining like terms

Remember, in algebra, a *term* is an expression that consists of numbers, fractions, and/or variables that are multiplied. This means that the expression $-2y + 7 + 8y$ has three terms, separated by the plus signs.

In the expression $-2y + 7 + 8y$, the terms $-2y$ and $8y$ are called **like terms** because they have the same variable part (in this case a single y). We can **combine** (add or subtract) like terms.

To do that, it helps to organize the terms in the expression in alphabetical order according to the variable part and write the constant terms last. We get $-2y + 8y + 7$ ($8y - 2y + 7$ is correct, too).

Next, we add $-2y + 8y$ and get $6y$. So the expression $-2y + 7 + 8y$ simplifies to $6y + 7$.

Example 3. Simplify $6y - 8 - 9y + 2 - 7y$.

First, we organize the expression so that the terms with y are written first, followed by the constant terms.

For that purpose, we **view each operation symbol (+ or −) in front of the term as the *sign* of each term.** In a sense, you can imagine each plus or minus symbol as being "glued" to the term that follows it. Of course the first term, $6y$, gets a "+" sign.

Why can we do it this way?

Because subtracting a term is the same as adding its opposite. In symbols,

$6y \quad -8 \quad -9y + 2 \quad -7y$
$= 6y + (-8) + (-9y) + 2 + (-7y)$.

In other words, the expression $6y - 8 - 9y + 2 - 7y$ is the SUM of the terms $6y$, -8, $-9y$, 2, and $-7y$.

After reordering the terms, the expression becomes $6y - 9y - 7y - 8 + 2$.

Now we need to combine the like terms $6y$, $-9y$, and $-7y$. We do that by finding the sum of their coefficients 6, −9, and −7. Since $6 - 9 - 7 = -10$, we know that $6y - 9y - 7y = -10y$.

Similarly, we combine the two constant terms: $-8 + 2 = -6$.

Our expression therefore simplifies to $-10y - 6$.

3. Fill in the pyramid! Add each pair of terms in neighboring blocks and write its sum in the block above it.

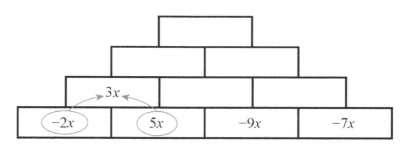

4. Organize the expressions so that the variable terms are written first, followed by constant terms.

   a. $6 + 2x - 3x - 7 + 11$

   b. $-s - 12 + 15s + 9 - 7s$

   c. $-8 + 5t - 2 - 6t$

5. Simplify the expressions in the previous exercise.

6. Simplify.

a. $5x - 8 - 7x + 1$	b. $-6a - 15a + 9a + 7a$
c. $-8 + 7c - 11c + 8 - c$	d. $10 - 5x - 8x - 9 + x$

7. Fill in the pyramid! Add each pair of terms in neighboring blocks and write its sum in the block above it.

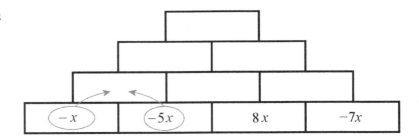

8. Find what is missing from the sums.

   a. $8x + 2 +$ _____ $= 5x + 8$

   b. $5b - 2 +$ _____ $= 2b + 7$

   c. $-2z +$ _____ $= 1 - 5z$

   d. $-4f + 3 +$ _____ $= -f - 1$

9. Fill in the pyramid! Add each pair of terms in neighboring blocks and write its sum in the block above it.

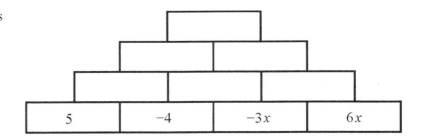

10. Simplify.

a. $0.5y + 1.2y - 0.6y$	b. $-1.6v - 1 - v$	c. $-0.8k + 3 + 0.9k$

11. A challenge! Solve the equation $(-1/2)x - 6 + 8x + 7 - x = 0$.

Example 4. One or both sides of an equation may have several terms with the unknown. In that case, we need to combine the like terms (simplify) before continuing with the actual solution.

$3x + 7 - 5x = 6x + 1 - 5x$	On the left side, combine $3x$ and $-5x$. On the right side, combine $6x$ and $-5x$.
$-2x + 7 = x + 1$	$-x$
$-3x + 7 = 1$	-7
$-3x = -6$	$\cdot (-1)$
$3x = 6$	$\div 3$
$x = 2$	

12. Solve. Check your solutions.

a. $\quad 6x + 3x + 1 = 9x - 2x - 7$

b. $\quad 16y - 4y - 3 = -4y - y$

c. $\quad -26x + 12x = -18x + 8x - 6$

d. $\quad -9h + 4h + 7 = -2 + 5h + 9h + 8h$

13. Solve. Check your solutions.

a. $2x - 4 - 7x = -8x + 5 + 2x$

b. $-6 - 4z - 3z = 5z + 8 - z$

c. $8 - 2m + 5m - 8m = 20 - m + 5m - 2m$

d. $-x - x + 2x = 5 - 5x + 9x$

e. $-q + 2q - 5q - 6q = 20 - 7 - 9 + q$

f. $9 - s + 7 - 9s = 2 - 2s - 11$

Some Problem Solving

Ratio problems and equations

Example 1. A truck is carrying 21 pallets of beans. Each pallet weighs 2,400 pounds and contains bags of pinto beans, navy beans, and black beans in a ratio of 7:3:2. Find the *total* weight of each type of beans in the truck.

We can solve this using a bar model. The entire bar signifies <u>one pallet</u>.

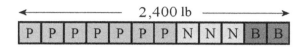

There are 7 + 3 + 2 = 12 parts in total. The weight of one part is 2,400 lb ÷ 12 = 200 lb.

Or we can write an equation. The unknown x is the same as in the bar model: it is one part.

$$7x + 3x + 2x = 2,400$$
$$12x = 2,400 \quad | \div 12$$
$$x = 200$$

But the weight of one part is not the final answer! We need to calculate the total weight of each type of beans:

- pinto beans: 21 · 7 · 200 lb = 29,400 lb
- navy beans: 21 · 3 · 200 lb = 12,600 lb
- black beans: 21 · 2 · 200 lb = 8,400 lb

1. Jaime and Juan divided a profit of $820 in a ratio of 2:3. Calculate each one's part.

    Jaime:

    Juan:

2. The *aspect ratio* of a photograph (the ratio of its width to its height) is 4:7 and its perimeter is 66 cm. How long and how wide is the photograph?

    Solve this problem in two ways, and observe how the two solutions relate to each other.

    a. Solve the problem using logical reasoning.

    b. Solve the problem using an equation.

3. You bought five computer keyboards from an online store. A shipping charge of $8.75 is included in the final bill, which comes to $78.20. How much did one keyboard cost?

   a. Solve the problem using logical reasoning.

   b. Write an equation for the problem, and solve it. Notice how the solution steps of the equation correspond to the steps of solving the problem using logical thinking.

4. Computer mice are on sale for $12 each from an online store. If your order is under $250, the store charges a shipping and handling fee of $7.90. You buy as many mice as you can with $100. How many mice can you buy?

   a. Solve the problem using logical reasoning.

   b. Write an equation for the problem, and solve it. Notice how the solution steps of the equation correspond to the steps of solving the problem using logical thinking.

5. The perimeter of the rectangle is 76 cm.
   Find its length and width.

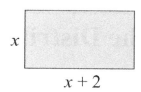

6. **a.** Sketch an equilateral triangle with a perimeter of $24x + 9$.

   b. Sketch a rectangle with a perimeter of $20x + 6$.

7. **a.** Write an equation to solve for the unknown a.

   b. Find the measure in degrees of each of the three angles.

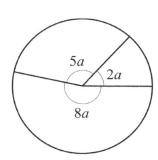

8. Four adjacent (side by side) angles form the line l.

   a. Write an equation to solve for the unknown a.

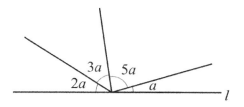

   b. Find the measure of each of the four angles, rounded to the nearest 0.01 degree.

Using The Distributive Property

Sometimes you may have to use the distributive property to remove parentheses before solving an equation.

Example 1.
$$2(x + 9) = -2x$$
$$2x + 18 = -2x \quad | +2x$$
$$4x + 18 = 0 \quad | -18$$
$$4x = -18 \quad | \div 4$$
$$x = -4.5$$

1. Solve.

a. $5(x + 2) = 85$

b. $9(y - 2) = 6y$

c. $2(x + 7) = 6x - 11$

d. $-3z = 5(z + 9)$

e. $10(x + 9) = -x - 5$

f. $2(5 - s) = 3s - 1$

Example 2. Sometimes there are two different ways that you can start to solve an equation.

First divide by a common factor:	$3(x+5) = -18$ $\quad$ ÷ 3 $x + 5 = -6$ $\quad$ − 5 $x = -11$	**First distribute the multiplication:**	$3(x+5) = -18$ $3x + 15 = -18$ $\quad$ − 15 $3x = -33$ $\quad$ ÷ 3 $x = -11$

2. Solve in two ways: (i) by dividing first and (ii) by distributing the multiplication over the parentheses first.

a. $6(x - 7) = 72$

Divide first:	**Distribute the multiplication first:**

b. $10(q - 5) = -60$

Divide first:	**Distribute the multiplication first:**

3. What is to be done in each step? Fill in the missing operations to be done to both sides of the equation.

| a. | $2(x + 9) = -2x$
 $x + 9 = -x$
 $2x + 9 = 0$
 $2x = -9$
 $x \quad -4\ 1/2$ | | b. | $\dfrac{x + 2}{7} = 6x$
 $x + 2 = 42x$
 $2 = 41x$
 $41x = 2$
 $x = 2/41$ |
 Switch the sides.
  |

4. Solve.

a.	$2(x + 7) = 3(x - 6)$	b.	$5(y - 2) + 6 = 9$
c.	$2x - 1 = 3(x - 1)$	d.	$3(w - ½) = 6(w + ½)$
e.	$20(x - ¼) = x - 2$	f.	$8 + 2(7 - v) = 13$

5. Select from the expressions at the right to make an equation that has -3 as its root.

$-x + 5$ $x - 5$ $4x$
12 $2(x + 4)$
$6(x + 1)$ $2(x - 4)$ $6(x - 1)$

6. Write an equation for this number line diagram, and find the value of the unknown *a*.

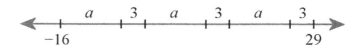

7. Draw a number line diagram for each equation, and solve.

   a. $-5 + 3x = 16$

   b. $-5 + 3(x + 2) = 16$

The distributive property works the same way with negative numbers as with positive ones. As you practice more, you'll be able to start doing these multiplications in a single step instead of several.

Example 3.
$-2(3x + 7) = -2 \cdot 3x + (-2) \cdot 7$
$= -6x + (-14)$
$= -6x - 14$

We write $+ (-14)$ as $- 14$, because that is simpler.

Example 4.
$-9(3s - 3)$
$= -9 \cdot 3s - 9(-3)$
$= -27s - (-27)$
$= -27s + 27$

The expression $9(-3)$ means $9 \cdot (-3)$. Similarly, $9 \cdot 3$ could be written as $(9)(3)$.

8. Use the distributive property to multiply.

a. $-11(x + 2) =$	b. $-3(y + 5) =$	c. $-7(x - 2) =$
d. $-3(y - 20) =$	e. $-10(2x + 10) =$	f. $-3(2y + 90) =$
g. $-0.5(w + 16) =$	h. $-0.8(5x - 20) =$	i. $-200(0.9x + 0.4) =$

9. Solve.

a. $-2(x + 5) = 5x$	b. $-6(y - 2) = 3(y + 2)$
c. $-0.1(40y + 90) = 2$	d. $3 - t = -5(t - ½)$

194

Example 5. The expression $-(x + 5)$ simply means $-1(x + 5)$. We can simplify it using the distributive property (on the right) →

$-1(x + 5) = -1x + (-1)(5)$
$= -x + (-5)$
$= -x - 5$

This is an example of an important principle in algebra:

If you **multiply any number by −1**, you get the opposite of that number. In symbols,

$$-1x = -x$$

So the expression $-(x + 5)$ means the *opposite* of $x + 5$. To simplify it, we take the opposite of each term in the quantity (the opposite of x *and* the opposite of 5) to get $-x - 5$.

Example 6. Simplify $-(2s - 5)$.

We multiply both terms in the quantity $2s - 5$ by -1. This ends up **changing the sign of each term in the quantity**, because each term becomes its opposite when multiplied by -1. So, both $2s$ and -5 change their signs and become $-2s$ and $+5$.

$-(2s - 5) =$
$(-1)2s - (-1)5$
$= -2s + 5$

10. Simplify.

a. $-(x + 4) =$	**b.** $-(y - 9) =$	**c.** $-(x - 12) =$
d. $-3(a - 6) =$	**e.** $2(-x + 5) =$	**f.** $-4(x - 1.2) =$
g. $-(4y - 20) =$	**h.** $-(5t + 0.9) =$	**i.** $-(2y + 1.4) =$
j. $-7(0.2 - w) =$	**k.** $0.1(-60t - 7) =$	**l.** $-0.2(9y + 4) =$

11. Here's another riddle to discover by solving the equations.

A. $3(2x - 4) = 3x$	**E.** $-2(x + 4) = 11$	**A.** $-(x + 4) = 10 + x$
S. $-2x + 4 = 3(5 - x)$	**A.** $-0.1(50x - 5) = x$	**R.** $2 = -2(x + 4)$
L. $5x = -(-2x - 3)$	**D.** $-6(3x + 1) = 1 + x$	**Y.** $3(2x + 5) = 4x + 5 + x$

Where can you buy a ruler that is 3 feet long? At...

4		−10	−7	−5	−7/19		11	1/12	1	−9 ½

The root of each equation below is 2. Find each of the missing numbers.

a. ___ $(z - 7) = -10$

b. ___ $(x - 5) = 3x + 6$

c. ___ $(y + 8) = 2y - 6$

d. ___ $w = -6(w + 1)$

Puzzle Corner

Word Problems

1. Which equation matches the problem? Once you find it, solve the equation.

   a. Mrs. Hendrickson bought herself a cup of coffee that cost $3. She also bought ice cream cones that cost $2.20 each for each of her preschoolers. She paid for all of it with $25. How many ice cream cones did she buy?

$n(3 + 2.20) = 25$	$3 + 2.20n = 25$
$3(n + 2.20) = 25$	$3n + 2.20 = 25$

   b. Mr. Sanchez spent about $35 to treat some people in his bicycling club to a cup of coffee and an ice cream cone each. Each coffee cost $3, and each ice cream cone cost $2.83. How many people were treated to coffee and ice cream by Mr. Sanchez?

$n(3 + 2.83) = 35$	$3 + 2.83n = 35$
$3(n + 2.83) = 35$	$3n + 2.83 = 35$

2. Write an equation for the following problem and solve it.
   The perimeter of a rectangle is 144 cm. Its length is 28 cm. What is its width?

3. An online store sells herbal sleeping aids for $12 per bottle. A fixed shipping and handling (s&h) fee of $5 is added to each order. (The s&h fee is not per bottle, but per order.)

   a. Write an expression for the total cost of buying five bottles.

   b. Write an expression for the total cost of buying n bottles.

   c. Write an equation for this question, and solve it.
      How many bottles can Mrs. Brooks get with $173?

4. Let's change the shipping and handling fee to be more realistic. Each bottle still costs $12, but the shipping and handling fee is $5 for the first bottle you buy, and $1 for each additional bottle in the same order.

   a. Write an expression for the total cost of buying five bottles.

   b. Write an expression for the total cost of buying eight bottles.

   c. Write an expression for the total cost of buying n bottles (where n is more than 1).

   d. Write an equation for this question, and solve it.
      How many bottles can Mrs. Brooks get with $173?

5. Based on the cost in problem 4, how many bottles can Mrs. Brooks buy with $300?

   a. Solve the problem without an equation.

   b. Solve this problem using an equation. Compare the solution steps to how you solved it in (a).

6. The total area of a sandbox with two compartments is 8 square meters. Solve for x.

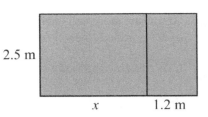

7. Before moving to a different suburb, Tony used to commute 14 km from his home to his workplace every workday. After his move, he was going to measure the new distance to his workplace using his car's odometer but forgot to check it until he had gone to work and back home three times. By then it showed 123.6 km.

    How much longer is Tony's commute now than it was in the past?

    a. Solve the problem using logical reasoning and arithmetic.

    b. Write an equation for the problem, and solve it.

8. A school schedule has six class periods of equal length plus a 30-minute study hall each day. The total time each day that students spend in classes and study hall is 360 minutes. How long is each class period?

    a. Solve the problem using logical reasoning and arithmetic.

    b. Solve the problem by writing an equation. Compare the steps of solving the equation to those in your method in part (a).

Inequalities

An **inequality** contains two expressions that are separated by one of these signs $<$, $>$, $\leq$ or $\geq$.

(expression 1) < (expression 2)

The sign $\leq$ is read, "less than or equal to." It is the $<$ sign and $=$ sign combined.
The sign $\geq$ is read, "greater than or equal to." It is the $>$ sign and $=$ sign combined.

For example, the inequality $6y \geq -2$ is read, "$6y$ is greater than or equal to a negative two."

Typically, an inequality has an infinite number of solutions. The set of all the possible solutions to an inequality is its **solution set**. We can **plot the solution set of an inequality on a number line.**

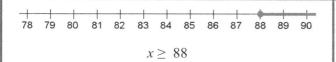

$x \geq 88$

We draw a *closed* circle at 88 because 88 fulfills the inequality. Any number that is greater than 88 (such as 88.1 or 90 1/2) also fulfills the inequality.

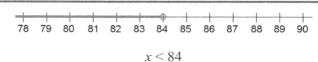

$x < 84$

We draw an *open* circle at 84, because 84 *does not* fulfill the inequality (84 < 84 is a false statement). Any number that is less than 84 works fine!

To solve an inequality, we can

- add the same number to both sides
- subtract the same number from both sides
- multiply or divide both sides by the same number.

So solving inequalities works essentially the same way as solving equations, but there is *one* exception. That is, if you divide or multiply an inequality by a *negative* number, you need to reverse the sign of the inequality (from $<$ to $>$, or $\leq$ to $\geq$, and vice versa). For example, multiplying the inequality $-7 < 7$ by -1 yields the inequality $7 > -7$. However, we will not be dealing with the exception in this course. You will be solving only inequalities where you multiply or divide the inequality by a *positive* number.

Example 1.
$5x + 3 < 6 \quad | -3$
$5x < 3 \quad | \div 5$
$x < 3/5$

Example 2.
$\dfrac{t - 20}{7} \geq 7 \quad | \cdot 7$
$\dfrac{\cancel{7} \cdot (t - 20)}{\cancel{7}} \geq 49$
$t - 20 \geq 49 \quad | +20$
$t \geq 69$

1. Write an inequality that corresponds to the plot on the number line.

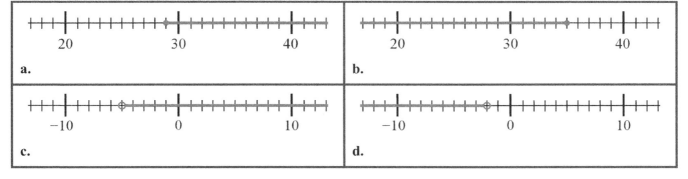

> **Example 3.** Mom said, "Don't spend more than $100." "Not more than" means "less than or equal to." You can spend $100 or any amount less than $100, but you cannot spend $100.29. We can represent this statement as
>
> $$\text{money spent} \leq \$100$$
>
> Using the variable m for the "money spent," we can write the inequality $m \leq \$100$.

> The symbol $\geq$ (greater than or equal to) often corresponds to the phrase "at least."
>
> "Could you get me at least 20 apples?" Using n for the number of apples, we can write $n \geq 20$.

2. Write an inequality for each phrase. Choose a variable to represent the quantity in question.

    a. You have to be at least 21 years of age.

    b. He has been to at least a dozen countries.

    c. Citizens 55 years or older get a free entry.

    d. We did not see more than 12 birds.

    e. I need at least 50 screws for the project.

3. Make up a situation from real life that could be described by the given inequality.

    a. $a < 30$

    b. $p > 100$

    c. $b \geq 8$

    d. $x \leq 60$

> **Example 4.** Solve the inequality $4x - 1 < 11$ in the set $\{1, 2, 5, 7, 9, 11\}$.
>
> This simply means to try each number from the set to see if it fulfills the inequality. For example, let's try 11. Is $4 \cdot 11 - 1$ less than 11? No, it's not (it's 43). So 11 is not a solution.
>
> In this case, the numbers 1 and 2 are the only solutions to the inequality. The solution set of this inequality is thus $\{1, 2\}$.

4. **a.** Solve the inequality $7x - 13 < 45$ in the set $\{1, 3, 5, 7, 9, 11\}$.

    b. Solve the inequality $3x + 10 \geq 25$ in the set $\{2, 4, 6, 8, 10\}$.

    c. Solve the inequality $2 - y \geq y + 1$ in the set $\{-3, -2, -1, 0, 1, 2, 3\}$.

5. Solve these inequalities by applying the same operation to both sides. Notice that the inequality symbol ($<, >, \leq$ or $\geq$) does not change.

a. $\quad 3y < 48$	b. $\quad y - 8 > 59$	c. $\quad 2c - 5 \geq 23$
$<$	$>$	$\geq$
		$\geq$

6. Solve each inequality and plot its solution set on the number line. For most of these, you need to write appropriate multiples of ten under the bolded tick marks (for example, 30, 40, and 50).

a. $\quad 2x + 12 < 30$	b. $\quad 3x - 5 > 83$
	$>$

c. $\quad 6x - 25 \leq 47$	d. $\quad 171 \geq 20x - 9$
$\leq$	$\geq$

e. $\quad 11a + 5 \leq 2a + 12$	f. $\quad 6 + 25y \geq 10y - 9$
$\leq$	$\geq$

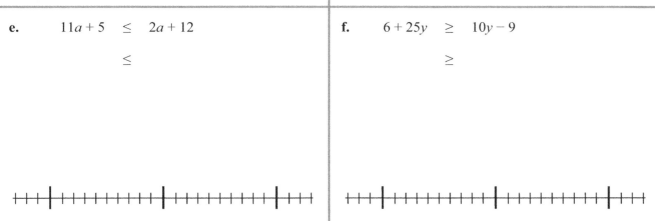

> **How do check the solution to an inequality?** After all, there is an infinite number of solutions.
>
> To check the solution to an inequality, choose some numbers from the solution set and check that they do indeed fulfill the inequality. Then choose some numbers that are *not* in the solution set and check that they *don't* fulfill the inequality.
>
> **Example.** $\quad 6x - 2 > 3 \qquad | +2$
>
> $\qquad\qquad\qquad 6x > 5 \qquad\quad | \div 6$
>
> $\qquad\qquad\qquad x > 5/6$
>
> To check that the solution $x > 5/6$ is correct, choose a few numbers that are greater than 5/6 and substitute them into the original inequality. It is best to choose at least one number that is near 5/6. Let's choose 1 and 2. The inequality should be true.
>
> $\quad 6 \cdot 1 - 2 \overset{?}{>} 3 \qquad\qquad\qquad\qquad 6 \cdot 2 - 2 \overset{?}{>} 3$
>
> $\qquad 4 > 3 \;\checkmark \qquad\qquad\qquad\qquad\quad 10 > 3 \;\checkmark$
>
> Then, we choose some numbers that are *not* in the solution set, say 0 and −5. This time, the inequality should NOT hold true.
>
> $\quad 6 \cdot 0 - 2 \overset{?}{>} 3 \qquad\qquad\qquad\qquad 6 \cdot (-5) - 2 \overset{?}{>} 3$
>
> $\qquad -2 > 3 \;\times \qquad\qquad\qquad\qquad\; -32 > 3 \;\times$
>
> Our check is complete and everything looks fine!

7. Solve the inequalities. Check your solution using the method above.

a. $\quad 2x - 3 > -9$	**b.** $\quad 9 + 3x \leq 30$
c. $\quad 5 + 10x < 22$	**d.** $\quad -20 + 3x \leq 19$

8. Which inequality has its solution plotted on the number line?

a. (i) $7x - 5 < 3$

   (ii) $5x - 3 < 7$

   (iii) $3x - 7 < 5$

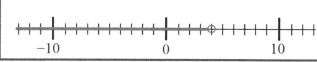

b. (i) $4x - 10 \le -34$

   (ii) $4x + 10 \ge -34$

   (iii) $4x - 10 \ge -34$

9. Solve the inequalities. Plot the solution set on the number line. Do not forget to check your answers.

a. $5x - 17 < 43$

b. $20x + 18 > 258$

The set of **natural numbers** or **counting numbers** is {1, 2, 3, 4, 5, 6...}.

10. **a.** What solutions does the inequality $x < 6$ have in the set of natural numbers?

   b. What solutions does the inequality $y > 11$ have in the set of *even* natural numbers?

   c. Find the natural numbers n that fulfill both inequalities: $n > 12$ and $n + 1 < 21$

11. Find three integers that fulfill both inequalities: $15y - 12 < 20$ and $2y + 6 > -5$.

Puzzle Corner

Find a number to fill in the empty space so that the solution to the inequality  $x - 7 < 23$ will be $x < 5$.

Word Problems and Inequalities

Example 1. As a salesperson, you are paid $150 per week plus $12 per sale. This week you want to earn at least $250. Write an inequality for the number of sales you need to make, solve it, and describe the solutions.

Let n be the number of sales, as that is the unknown. Each week, you earn $\$150 + \$12 \cdot n$. Let's write this expression without the dollar symbols and in the usual order of terms: $12n + 150$.

That expression, with the condition that you want to earn at least $250, gives the inequality $12n + 150 \geq 250$.

$$12n + 150 \geq 250 \quad \Big| -150$$

$$12n \geq 100 \quad \Big| \div 12$$

$$n \geq 8\,4/12$$

$$n \geq 8\,1/3$$

Since n is the number of sales, it has to be a whole number. Therefore, if you make 9 or more sales, you will be paid at least $250.

Another way to solve this is to think in terms of the amount you want to earn on top of your $150 base pay. In that case, you can write the inequality $12n \geq 100$.

1. Let n represent the amount of sales you make as a salesperson. Which description(s) of the solution set are correct?

a. $n < 18$	b. $n \leq 30$
You make at least 18 sales.	You make at least 30 sales.
You make at most 17 sales.	You make 29 sales or fewer.
You make at most 18 sales.	You make at most 30 sales.
You make 19 sales or more.	You make 31 sales or more.

2. Jeannie earns $350 per week plus $18 for each hour of overtime that she works. How many hours of overtime does she need to work if she wants to earn at least $500?

Write an inequality and solve it. Plot the solution set on a number line.
Lastly, explain the solution set in words.

*Write an inequality for each problem, and solve it. Plot the solution set on a number line.
Lastly, explain the solution set in words.*

3. Sheila is packing decorative candles for shipping. Each candle comes in its own box, which is 6.5 cm tall. The carton for shipping them is 45 cm tall. If she begins packing by putting a 1-cm-thick pad in the bottom, then how many candle boxes will she be able to stack vertically in the shipping carton?

4. You have a coupon that gives you a $10 discount on your total order at SuperSales store. How many pairs of socks can you buy, if you have only $55 in cash to spend, and the socks cost $6.70 a pair?

5. Joey has to work a minimum of 160 hours as an apprentice in a print shop before he can be hired permanently. He has already completed 21 hours of training. Now he has an opportunity to work 7.5-hour workdays. How many days will he need to work in order to finish at least the minimum training?

6. You want to print an ebook at a local office center. Printing black and white pages costs $0.03 per page, and then there is a fixed binding fee of $2.45.

   How many pages can you print for no more than $35?

Graphing

Do you remember equations with two variables? When an equation has two variables (like the equation $y = 2x - 3$), it usually has an infinite number of solutions. In other words, there is an infinite number of values for x and y that make the equation true.

For example, if $x = 0$, then we can calculate the value of y using the equation: $y = 2 \cdot 0 - 3 = -3$. So when $x = 0$ and $y = -3$, the equation is true. The **number pair** $(x, y) = (0, -3)$ is a solution.

Similarly, if x is 3, then $y = 2 \cdot 3 - 3 = 3$. The number pair $(3, 3)$ is *also* a solution.

This way, we could generate an infinite number of solutions. Each solution is a number pair that can be plotted on a coordinate grid.

This table lists some x and y values, plotted at the right, for the equation $y = 2x - 3$:

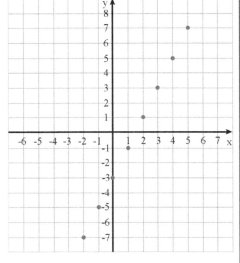

x	−2	−1	0	1	2	3	4	5
y	−7	−5	−3	−1	1	3	5	7

Notice the pattern in the table and in the graph: as the x-values increase by 1, the y-values increase by 2. The plot shows a pattern, as well: the dots form a line that is rising upwards.

1. Plot the points from the equations for the values of x in the table. Graph both (a) and (b) in the same grid.

a. $y = x + 4$

x	−9	−8	−7	−6	−5	−4	−3	−2
y								

x	−1	0	1	2	3	4	5
y							

b. $y = 2x - 1$

x	−3	−2	−1	0	1	2	3	4	5
y									

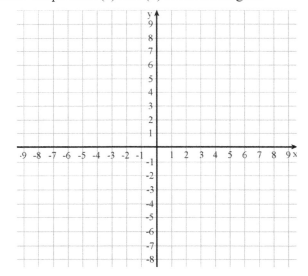

2. Which equation matches the plot on the right?

$y = (½)x + 1$

$y = (½)x$

$y = (½)x - 1$

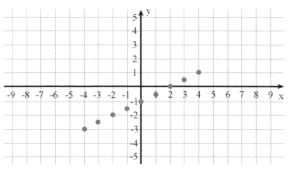

3. Plot the points from the equations for the values of x listed in the table. Graph both (a) and (b) in the same grid.

a. $y = -x + 2$

x	−5	−4	−3	−2	−1	0	1	2	3
y									

b. $y = -2x + 1$

x	−4	−3	−2	−1	0	1	2	3	4
y									

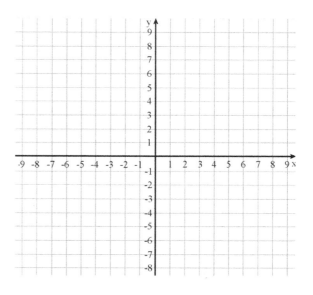

> **Example 1.** Is $(5, -2)$ a solution to the equation $y + 3 = 3x$?
>
> Simply substitute the values $x = 5$ and $y = -2$ into the equation, and check if you get a true or false equation (on the right).
>
> $-2 + 3 \stackrel{?}{=} 3 \cdot 5$
>
> $1 \stackrel{?}{=} 15$ ✗
>
> Be careful that you substitute the value of x in place of x in the equation, and not in place of y!
>
> No, $(5, -2)$ is not a solution to $y + 3 = 3x$.

4. **a.** Does $(-1, 0)$ fulfill the equation $y = 2x - 1$?

   b. Is $(2, -3)$ a solution to the equation $y - 1 = -x$?

5. Write an equation in the form "$y = mx + b$" (where m and b are constants) that shows how to calculate the value of y from the value of x. Graph the points if it helps you.

a.

x	−30	−20	−10	0	10	20	30
y	−20	−10	0	10	20	30	40

equation: _____

b.

x	−30	−20	−10	0	10	20	30
y	−60	−40	−20	0	20	40	60

equation: _____

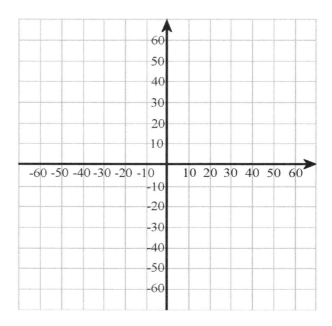

Frequently, we plot an equation for *all* values of x, not just for the integer values. This means x can be any decimal or fraction, for example 2.56 or 2 3/4. If x is a decimal, then y probably will be, too.

This would make for a tremendous amount of number pairs to be plotted as dots. We cannot draw that many dots on the grid, so instead we draw a smooth *line* (or a curve for some types of equations). **The line represents all of those specific number pairs (dots).**

Example 2. Graph the equation $y = -2x$.

First, we will plot several points that make the equation $y = -2x$ true, just like we did before.

How do we select those points? We simply choose any x values we like and calculate the corresponding y values. For example:

x	−3	−2	−1	0	1	2	3
y	6	4	2	0	−2	−4	−6

We plot those points and see that they fall on a line.

Lastly, we draw a line through the points that extends as far as it can go in both directions. This line now represents all the points that fulfill the equation—even the ones with fractional and decimal coordinates.

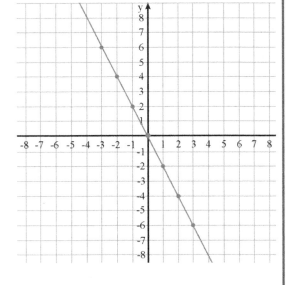

Since the plot of this equation is a line, we call the equation a **linear equation**. You can learn more about them in an algebra course.

Graph the equations (as lines). Graph two equations in each coordinate grid.

6. **a.** $y = x + 2$ **c.** $y = (½)x + 2$

   b. $y = -x + 3$ **d.** $y = 2x - 2$

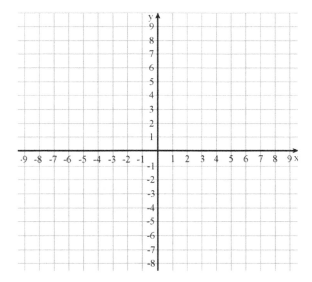 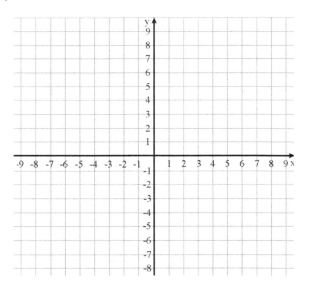

7. How would you check if a given point is on a given line without drawing anything?
For example, is the point (1, −2) on the line $y = x - 2$?

8. Match each equation with its graph.

   $y = (1/2)x - 3$

   $y = 4 - x$

   $y = -2x + 3$

   $y = (2/3)x - 1$

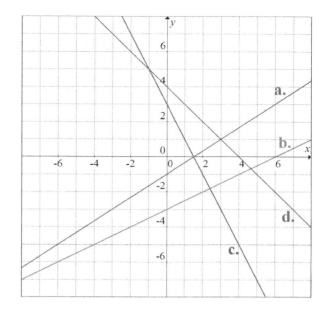

9. Is the point $(-1, 1)$ on the line $y = x + 1$?

10. **a.** Plot the equation $y = (1/2)x - 2$.

    b. Plot the equation $y = -3x$.

    c. Plot the equation $y = 6 - x$.

11. Explain two different ways to determine if the point $(5, -5)$ is on the line $y = -(1/2)x - 2$.

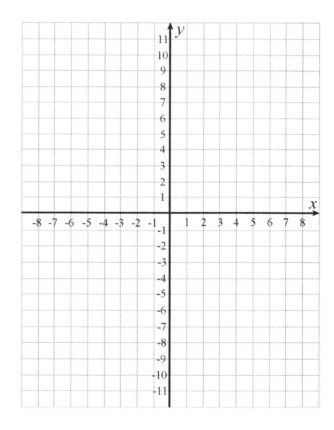

Puzzle Corner

x	−3	−2	−1	0	1	2	3
y	−5	−3	−1	1	3	5	7

Write an equation that relates x and y.

Hint: It is of the form $y = mx + b$ where m and b are integers.

An Introduction to Slope

The **slope** of a line is a number that describes the steepness and direction of its slant or inclination. It is defined as **how many units the *y*-value changes (the "rise") when the *x*-value is changed (the "run") by 1 unit**.

Example 1. What is the slope of the line $y = 2x - 1$?

Looking at the table of some *x* and *y* values, we see that every time the *x*-coordinate increases by 1 unit, the *y*-coordinate increases by 2 units. This means the slope is 2.

x	−4	−3	−2	−1	0	1	2
y	−9	−7	−5	−3	−1	1	3

From any point on the line, draw a one-unit horizontal line segment (the "run") toward the positive *x* direction (right). From the end of that segment, draw another line segment (the "rise") toward the positive *y* direction (up) until you meet the line again. How long is that vertical segment? It is 2 units long. So the slope — the "rise" per unit "run" — is 2.

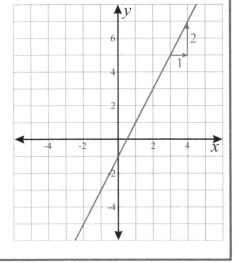

1. Find the slope of the lines. You can use the table, plot the line, or do both.

a. $y = 3x - 2$

x								
y								

b. $y = x + 5$

x								
y								

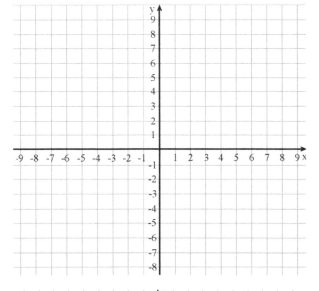

c. $y = (1/2)x$

x								
y								

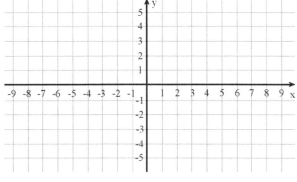

The slope can also be a negative number. In that case, we can say that the line is **decreasing**. Moving from left to right, it goes downwards. Conversely, if the slope is positive, the line is **increasing**.

Example 2. What is the slope of the line $y = -x + 3$?

Every time the *x*-coordinate increases by 1 unit, the *y*-coordinate *decreases* by 1 unit. This means the slope is −1.

x	−4	−3	−2	−1	0	1	2
y	7	6	5	4	3	2	1

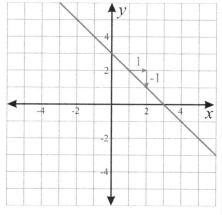

To determine the slope from the graph, again pick any point on the line and draw a one-unit horizontal line segment toward positive *x*. This time to meet the line you have to draw the second segment downwards. The vertical distance is 1 unit, but the "rise" is negative, toward negative *y*, so the slope is −1.

2. Find the slope of each line. You can use the table, plot the line, or do both.

a. $y = -3x + 1$

x							
y							

b. $y = -2x$

x							
y							

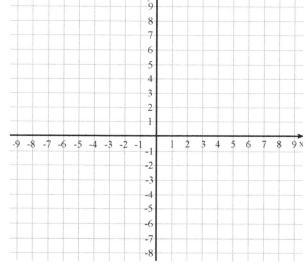

c. $y = 2x - 2$

x							
y							

d. $y = -(1/2)x + 4$

x							
y							

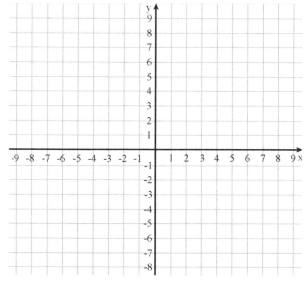

Sometimes it is easier to use an increase of some amount other than 1 unit in the x-coordinate. We still use the corresponding change (increase or decrease) in the y-coordinate, but the slope is the ratio of the two changes. So, another way to calculate the slope is:

$$\text{slope} = \frac{\text{change in } y\text{-coordinates}}{\text{change in } x\text{-coordinates}} \quad \text{or} \quad \text{slope} = \frac{\text{rise}}{\text{run}}$$

This is often expressed as "rise over run." The **rise** is the change in the y-coordinates—the change in the vertical direction. The **run** is the change in the x-coordinates—the change in the horizontal direction.

Example 3. Determine the slope from the graph.

As the x-coordinates increase by 3 units (the run), the y-coordinates increase by 5 units (the rise). So the slope is 5/3.

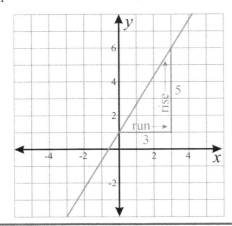

Example 4. Determine the slope from the table.

We can see from the table that each time the x-coordinates increase by 10 units (the run), the y-coordinates *decrease* by 4 units (the rise). Since the rise is negative, so is the slope.

The slope is the ratio of the two changes: *rise / run* = −4/10 or −2/5.

x	0	10	20	30	40	50	60
y	16	12	8	4	0	−4	−8

3. Determine the slopes from the graphs. Remember that for a decreasing line, the change in the y-coordinates is negative, which makes the slope negative.

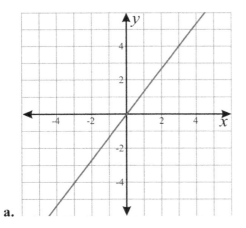

a.

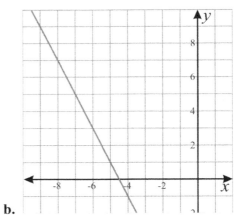

b.

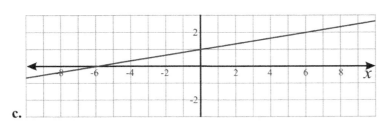

c.

4. Determine the slope of each line from the table or from its graph.

   a.
x	−3	−2	−1	0	1	2	3
y	−3 ½	−2	−½	1	2 ½	4	5 ½

   b.
x	−4	−2	0	2	4	6	8
y	5	4	3	2	1	0	−1

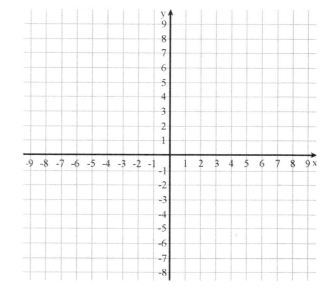

5. Determine the slope of each line. The scaling of the grids is different from that of the grids you have seen in this lesson so far, but the way to find the slope is the same: rise over run.

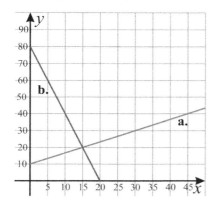 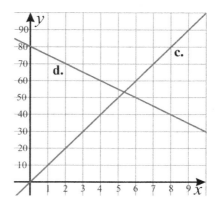

6. Draw two lines with a slope of 3/4. They can be drawn anywhere on the grid; they do not have to go through any specific point.

   Check: Your lines should be parallel.

7. Draw two lines with a slope of −3/4.

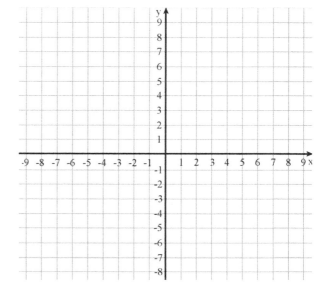

8. **a.** Draw a line that has a slope of 1/2 and that goes through the point (0, 6).
   Hint: Start at the point (0, 6), and draw the "rise/run diagram" using the ratio 1:2.

   b. Draw a line that has a slope of −3 and that goes through the point (−5, 6).

   c. Draw a line that has a slope of 2/3 and that goes through the point (0, 1).

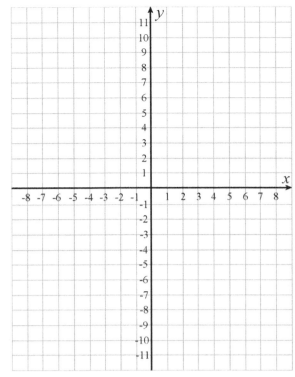

9. Draw any line with a slope of 30.

10. Draw a line that goes through the point (1, 70) and has a slope of −15.

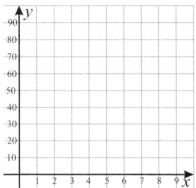

11. **a.** What is the slope of a line that goes through the points (2, 5) and (3, 8)?

    b. What is the slope of a line that goes through the points (6, 9) and (7, 3)?

12. Sean tried to determine the slope of the line in this graph. He said, "The slope is 20, because the line goes through the point (10, 20)." Play teacher and explain what's wrong with Sean's reasoning, how he can find the correct answer, and what that answer is.

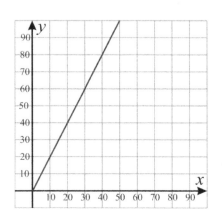

Speed, Time, and Distance

We studied the formula $d = vt$ in chapter 3. It tells us how the quantities *distance* (d), *velocity* (v), and *time* (t) are interrelated when an object travels at a constant speed. Their relationship can also be written as $v = d/t$, which you can derive from the common unit for speed, "kilometers per hour."

In this lesson, we explore the relationships between speed, time, and distance in the context of graphing.

Example 1. Harry runs along a 100-meter track at a constant speed. The table below shows his position or distance (d) from the starting line in relation to time (t).

t	0	1	2	3	4	5	6
d	0	5	10	15	20	25	30

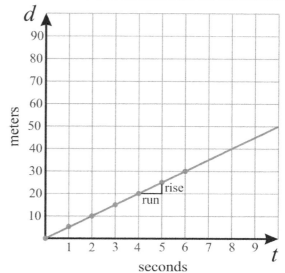

We graph the points and then draw a line through them.

Notice that for each second of time that passes, Harry advances 5 meters. This gives us the "rise/run" relationship that determines the slope of the line.

So, the slope is (5 m)/(1 s), or 5 meters per second. This slope, or change in position over time, is simply Harry's speed.

We can use the slope to relate the quantities t and d in a simple equation: $d = 5t$. Notice that this is simply the formula $d = vt$ with a velocity v of 5 m/s.

In reality, we have to express the velocity in some unit of measure (meters per second in this case), but when we write a formula or an equation, we usually omit the units as a convenience and simply write $d = 5t$ instead of $d = 5$ m/s $\cdot t$. However, you still need to include the units in your calculations and final answers.

1. Graph the points. Draw a line through them. Write an equation that relates t and d.

a.

t	0	1	2	3	4	5	6
d	0	4	8	12	16	20	24

equation: _____

b.

t	0	1	2	3	4	5	6
d	0	7	14	21	28	35	42

equation: _____

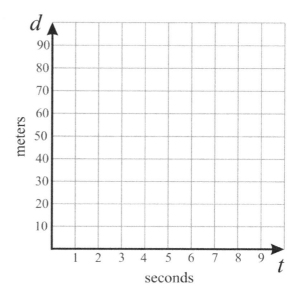

c. If the lines represent two runners running with a constant speed, how far from the starting line is each runner when $t = 12$ s?

2. The graph below shows how Henry ran the 100-meter dash.

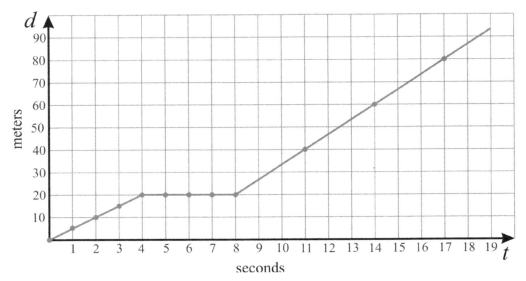

a. For the first four seconds, Henry runs at a constant speed.
What is his speed?

Also, write an equation relating distance and time in the first four seconds.

b. What happens from the time 4 seconds till 8 seconds?
Look also at the table on the right.
Notice that Henry's position does not change!

t	4	5	6	7	8
d	20	20	20	20	20

c. From 8 seconds and onward, Henry runs at a constant speed again, but it is different from his earlier speed. What is his speed now?

d. How can we tell *from the graph* that Henry ran at a different speed at the beginning than at the end?

e. Finish drawing the graph until Henry reaches the 100-meter line.
What is Henry's total running time for the 100-meter race?

3. Sally runs at a constant speed of 5 m/s for six seconds. Then she stops for six seconds to tie her shoelaces. Then she runs to the finish line (at 100 m) at a speed of 7.5 m/s.

   a. Plot a graph for the distance Sally runs.

   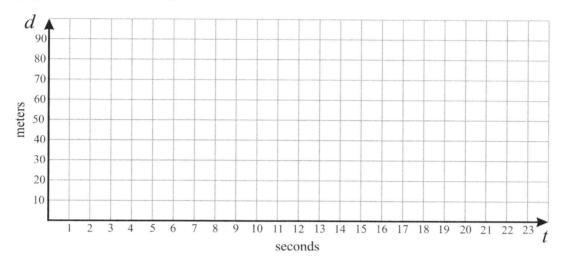

   b. What is Sally's total running time for the race?

4. Daisy starts the race, not at the 0-meter mark, but at the 10-meter mark. She runs at a speed of 4 m/s for the first eight seconds. Then she runs the rest of the way (to 100-meter mark) with the speed of 6 m/s.

   a. Plot a graph for the distance Daisy runs.

   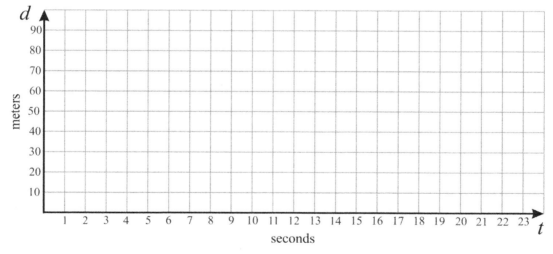

   b. How long does she take to run the race?

5. An airplane travels at a constant speed of 800 km/h from Phoenix to Chicago, a distance of 2,320 km.

   a. Write an equation relating the distance (d) it has traveled and the time (t) that has passed.

   b. Plot your equation.

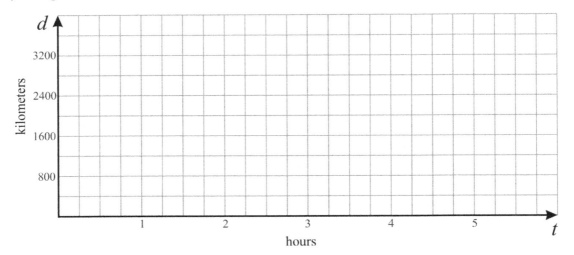

   c. How far will the airplane travel in 2 h 45 minutes?

   d. At what time will the airplane reach Chicago if it left Phoenix at 10:30 am?

6. Tom used a trencher at a steady rate. The graph shows the length of trench he dug in relation to the time he spent.

   a. What is Tom's trenching speed?
   Do not forget to include the units in your answer.

   b. What is the slope of the line in the graph?

   c. Write an equation relating distance (in feet) and time (in minutes).

   d. At the same rate, how many feet of trench could Tom dig in 2 hours?

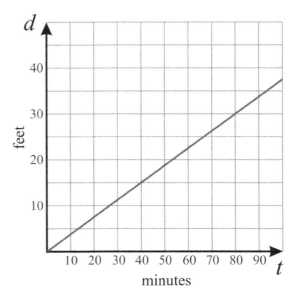

7. Mary drives her tractor at a constant speed. The equation $d = 25t$ tells us the distance, in kilometers, that she travels in t hours.

   a. What is Mary's speed, in kilometers per hour?
   Hint: Find how far Mary can travel in one hour.

   b. Plot the equation $d = 25t$.

   c. Jerry drove a tractor from his home to a nearby store, a distance of 3 km, in 10 minutes. What was his average speed (in kilometers per hour)?

   d. Let's say Jerry drives his tractor at the same constant speed as in (c). Write an equation relating the distance Jerry covers and the time it takes him, and plot it.

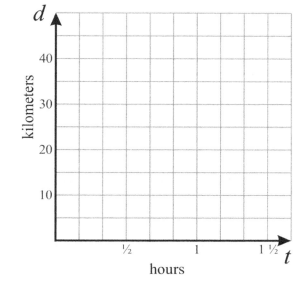

   e. Use the graph to estimate how much farther Mary can drive than Jerry in 50 minutes.

   f. Find the exact answer to (e).

Puzzle Corner

Joe starts the race at the 100-meter mark. He runs *towards the beginning* at a constant speed of 4 m/s until the 30-meter mark. Then he stops there for four seconds. Then he runs at a constant speed of 3 m/s to the starting line. Plot the distance between where Joe started to run and the zero-meter line in relation to time.

Chapter 5 Mixed Review

1. Find the missing numbers and terms.

| a. ____ $(6x - 5) = 72x - 60$ | b. $12($ ____ $-$ ____ $+$ ____ $) = 108y - 36x + 4.8$ |

2. Write an expression with two terms: the coefficient of the first term is 5, its variable part is x cubed, and the second term is the constant $-1/2$.

3. **a.** Which equation matches the situation?

   A town of population p lost 2/3 of its population, and now it has 2,600 residents.

   $p - 2/3 = 2600$ $\dfrac{2p}{3} = 2600$ $p - 1/3 = 2600$ $p - 2/3p = 2600$

   b. How many people lived in the town originally?

4. Add.

| a. $(-14) + 7 + (-8) + 2 =$ | b. $-3 + (-12) + 21 + (-19) + (-5) =$ |

5. Give a real-life context for each multiplication. Then solve.

| a. $1.4 \cdot 119$ |

| b. $(9/10) \cdot 14.30$ |

6. Change each subtraction into an addition, then add.

a. $-8 - (-7) - (-12) =$	b. $63 - (-11) + (-5) =$

7. a. Write an expression for the distance between x and 8.

   b. Evaluate your expression if $x = -52$.

8. Solve using both decimal and fraction arithmetic.

a. $0.24 \div 0.03$ **Decimal division:**   **Fraction division:**

b. $7.1 \cdot 0.5$ **Decimal multiplication:**   **Fraction multiplication:**

9. Solve.

$$\frac{5}{6} \cdot \frac{2}{3} \div \frac{4}{3}$$

10. Solve *without* a calculator.

a. 11% of $15	b. 90% of −12	c. 75% of −200 m

11. Solve *without* a calculator. Change the decimals into fractions or treat fractions as divisions.

a. $0.5 \cdot \frac{11}{12}$	b. $\frac{2}{5} \cdot (-0.8)$	c. $-\frac{5}{6} \cdot 0.2$

12. Rewrite each expression without parentheses.

a. $2 + (-g) =$	b. $15 - (-r) =$	c. $7x + (-2y) =$

13. Write the numbers in scientific notation.

   a. 113,000

   b. 45,980,000

14. Simplify the complex fractions.

a. $\dfrac{\frac{7}{8}}{\frac{8}{9}}$	b. $\dfrac{\frac{1}{2}}{\frac{1}{5}}$	c. $\dfrac{\frac{15}{21}}{\frac{2}{3}}$

Chapter 5 Review

1. Solve. Check your solutions (as always!).

a.	$1 - 3x = 17$	**b.**	$29 = -6 - 2y$
c.	$\dfrac{3x}{8} = 42$	**d.**	$\dfrac{v-2}{7} = -13$
e.	$\dfrac{w}{40} - 7 = 19$	**f.**	$\dfrac{s+8}{-3} = -1$

2. Solve each problem in two ways: (1) by writing an equation and (2) by using logical reasoning or a bar model.

a. You bought 15 bottles of oil for the equipment in your lawn-care business. You got a $14 discount on your entire purchase. The total cost was $130 after the discount. What is the normal price of one bottle of oil?	
Equation:	Logical thinking:

b. Three-sevenths of a number is 153. What is the number?	
Equation:	Logical thinking:

3. A carpet salesman earns a base salary of $300 a week. He also earns an additional $18 for every carpet he sells.

   a. Write an expression for the salesman's total weekly earnings if he sells n carpets.

   b. How many carpets does the salesman need to sell in order to earn $750 in a week? Write an equation and solve it.

4. Solve. Check your solutions.

a. $\quad 2x + 6 + 3x \;=\; 9x - 11$	**b.** $\quad 2(x + 6) \;=\; 9x - 11$
c. $\quad 6(5 - w) \;=\; 2(9 - w)$	**d.** $\quad -10(4y + 7) \;=\; -9y$

5. Four adjacent (side-by-side) angles form the line *l*.

   a. Write an equation to solve for the unknown *x*.

   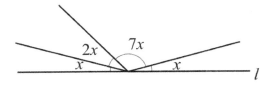

   b. Solve your equation and find the measure of each of the four angles.

6. The total area of a divided room is 200 square feet. Find the unknown dimension.

   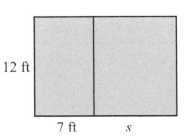

7. Solve the inequalities and plot their solution sets on the number line. You need to write appropriate numbers for the tick marks yourself.

a. $5x - 8 < 22$	b. $x + 5 \geq -2$

8. Write an equation for this number line diagram and solve it to find the value of the unknown y.

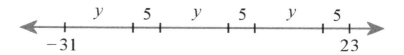

9. An airline has a weight limit of 20 kg for carry-on bags (the luggage passengers carry onto the airplane). Sharon's clothes, personal items, and the carry-on bag itself weigh 9 kg. Besides those, she wants to take a camera that weighs 2.6 kg and as many 0.8-kg bags of nuts as she can. How many bags of nuts can she take?

   a. Solve the problem without an equation or inequality.

   b. Write an inequality for the problem and solve it.

10. Find the slope of each line. Also, graph the lines.

   a. $y = -2x - 1$

x							
y							

   Slope: _____

   b.

x	−3	−2	−1	0	1	2	3
y	−6 ½	−4	−1 ½	1	3 ½	6	8 ½

   Slope: _____

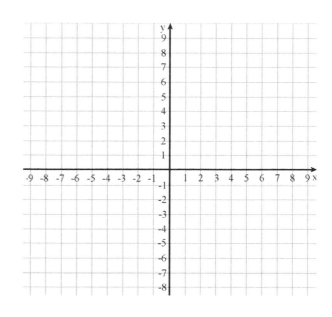

11. Draw a line with a slope of 5/6.

12. Draw a line that has a slope of 3/2 and that goes through the point (0, 2).

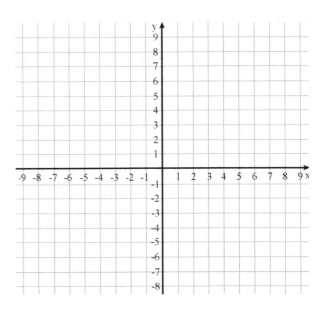

13. An airplane travels at a constant speed of 600 mi/h from New York to Los Angeles, a distance of 2,450 miles.

   a. Write an equation relating the distance (*d*) it has traveled and the time (*t*) that has passed.

   b. Plot your equation. Notice that you need to scale the *d*-axis.

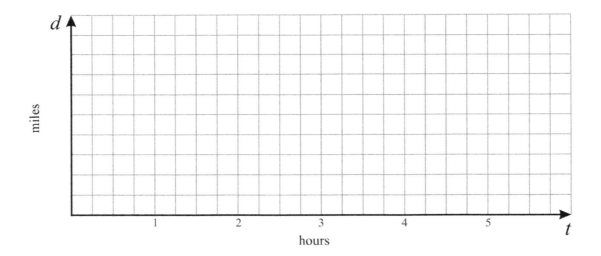

   c. How far will the airplane travel in 1 hour 40 minutes?

Made in United States
Orlando, FL
23 August 2023